THE APOCALYPSE:

COMETS, ASTEROIDS AND CYCLICAL CATASTROPHES

Laura Knight-Jadczyk

The Apocalypse:

Comets, Asteroids and Cyclical Catastrophes

Red Pill Press
www.redpillpress.com

First edition.
Copyright © 2012 Laura Knight-Jadczyk.
Research Sponsored by Quantum Future Group, Inc.
P. O. Box 5357 Baltimore, MD 21209

ISBN 978-1-897244-61-6

Design & illustrations: © 2012 Quantum Future Group, Inc., The Fellowship of
the Cosmic Mind, Inc., and Dreamstime LLC, except for: page 17 © Faulkes
Telescope South, and page 182 © Fox40.

Portions of this book were published 2010-2011 in *The Dot Connector Magazine* –
www.thedotconnector.org

Contents

Introduction

In this book, I will argue that there is a lot of evidence that our planet undergoes cataclysmic bombardment by comets and their fragments a lot more often than most scientists, scholars and the general public think or believe. As John Lewis, Professor of Planetary Sciences at the Lunar and Planetary Laboratory, Co-director of the NASA/University of Arizona Space Engineering Research Center, and Commissioner of the Arizona State Space Commission wrote in his book, *Comet and Asteroid Impact Hazards on a Populated Earth*:

> Awareness of the possibility of large impact events on Earth, although long present among a handful of the most imaginative thinkers, has come of age in this century as a result of studies of Arizona's Meteor Crater and the Tunguska fireball of June 30, 1908, in Siberia, spacecraft observations of cratering on Earth and other rocky bodies, and astronomical surveys of the near-Earth asteroid and comet populations. Appreciation of the effects of large impacts has developed in response to these studies and to the unclassified literature on the effects of large nuclear weapons.
>
> … [T]he most intensively studied impact phenomenon, impact cratering, is of limited importance, due to the rarity and large mean time between events for crater-forming impacts. Almost all events causing property damage and lethality are due to bodies less than 100 meters in diameter, almost all of which, except for the very largest and strongest, are fated to explode in the atmosphere. … Since explosions greater than 1 gigaton TNT are rare on this short of a time scale, we are forced to conclude that the complex behavior of smaller bodies is closely relevant to the threat actually experienced by contemporary civilization.
>
> … [T]he large majority of lethal events (not of the number of fatalities) are caused by bodies that are so small, so faint, and so numerous that the cost of the effort required to find, track, predict, and intercept them exceeds the cost of the damage incurred by ignoring them. (Lewis 1999, xi, xiii–xiv)

Lewis' statement above pretty much sums up the conclusions of all the research into comet and asteroid hazards that has been going on in a somewhat frenzied way for the past 14 years or so. The ones we really have to worry about, the ones that will kill people on the planet in random, unforeseen disasters, probably can't be seen and are too numerous for it to be cost effective to try to find and deal with them. In other words, the public is abandoned to their fate. *Not only abandoned, but a deliberate policy of concealment of the facts is clearly in place*, as we'll see.

Thumbing through *Comet/Asteroid Impacts and Human Society* (edited by Peter T. Bobrowsky and Hans Rickman), a collection of scientific papers presented at a workshop under the aegis of the International Council for Science and published by the eminent scientific publishing house, Springer, we read in the introduction: *"The International Council for Science recently recognized that the societal implications (social, cultural, political and economic) of a comet/asteroid impact on Earth warrants an immediate consideration by all countries in the world."* Pardon the sarcasm, but wow! You think? It seems that it's not just yours truly (and a few others on the Net) who are keeping track of the increasing number of fireballs and meteorites that suggest we are passing through rather dangerous areas of space, or that maybe 'something wicked this way comes'.

In the chapter entitled 'Social Perspectives on Comet/Asteroid Impact (CAI) Hazards: Technocratic Authority and the Geography of Social Vulnerability' from the above-cited book by Kenneth Hewitt, we read:

> Until quite recently, research into comet and asteroid hazards was focused on establishing the scale and scope of past impacts, credible estimates of their recurrence, and models for physical impact scenarios. ... CAI hazards have moved well beyond the realm of ungrounded speculation and apocalyptic visions. The results represent more than just new findings. They revolutionize, or are about to revolutionize, some basic understandings about the Earth, its history, biological evolution and future. Although human life has had a tiny place in the story so far, our longer term fate seems to be challenged by these forces and may be decided by them. (Bobrowsky & Rickman 2007, 399)

No kidding. Then, in a chapter entitled 'Social Science and Near-Earth Objects: an Inventory of Issues' by Lee Clarke, we read:

> It would have been ridiculous, not too long ago, to admit openly that you were thinking about asteroids and comets slamming into the Earth. Such events could mean the end of the world as we know it – TEOTWAWKI as millenialists call it – and that kind of talk is often ridiculed. ...
>
> Respectable people are pondering the issues. For example, S. Pete Worden, who is a Brigadier General in the US Air Force and Deputy Director for Command

and Control Headquarters at the Pentagon, has said that he believes "we should pay more attention to the 'Tunguska-class' objects – 100 meter or so objects which can strike up to several times per century with the destructiveness of a nuclear weapon" ... (Bobrowsky & Rickman 2007, 355)

I located the general's comments and it seems that the above is not all the general said. In fact, he states quite unequivocally:

I can show people evidence of real strikes inflicting local and regional damage less than a century ago. Even more compelling are the frequent kiloton-level detonations our early warning satellites see in the earth's atmosphere. ... Within the United States space community there is a growing concern over "space situational awareness."

The general was writing back in 2000. *"Less than a century ago."* That would be after 1900. In other words, he is saying that there were *"real strikes inflicting local and regional damage"* since 1900?! Did I miss something? Did all of us miss something? It seems so.

Several of my research helpers did a little digging on the question after reading the above quote and came up with some very interesting finds. It seems that there were two events in the 1930s that equaled the Tunguska event (described in detail in a later chapter). An article, 'Two "Tunguskas" in South America in the 1930's?' was printed in International Meteor Organization's December 1995 edition of the *WGN* journal.[1] It was written by Duncan Steel of the Anglo-Australian Observatory, who writes:

There is evidence that there were two massive bolide explosions which occurred over South America in the 1930's. One seems to have occurred over Amazonia, near the Brazil-Peru border, on August 13, 1930, whilst the other was over British Guyana on December 11, 1935. It is noted that these dates coincide with the peaks of the Perseids and the Geminids, although any association with those meteor showers is very tentative.

The identification of such events is significant in particular in that they point to the need for re-assessment of the frequency of Tunguska-type atmospheric detonations.

Then there is the article 'A rain of around 70 tons of Iron' by George Zay of Sky Publishing Corporation, who publish *Sky and Telescope Magazine*. Zay writes:

---

[1] http://www.xtec.cat/~aparra1/astronom/craters/amazonase.htm

This week marks the golden anniversary of what is arguably the most spectacular meteorite fall ever seen. At 10:40 a.m. on February 12, 1947, a incredibly bright fireball seared its way across the sky of eastern Siberia and rained around 70 tons of iron meteorites onto the rugged landscape. Because it was so well documented, the Sikhote-Alin fall proved a great boon to meteorite science.

The 1947 Siberian event is considered in most literature as one of the two most significant events this century where the earth has encountered objects from space. It was an iron meteorite that broke up only about 5 miles above the earth. It produced over 100 craters with the largest being around 85 feet in diameter. The strewn field covered an area of about 1 mile by a half mile. There were no fires or similar destruction like that found at Tunguska. Shredded trees and broken branches mostly. A total of 23 tons of meteorites were recovered and it's been estimated its total mass was around 70 tons when it broke up.[2]

There are more, of course, but this just tells us that there are many things going on here on the Big Blue Marble that we aren't aware of. That's exactly what Victor Clube was saying in his narrative report to the USAF and Oxford that sent me off on this topic in the first place. So let's get right to it and take a look at what Clube had to say that was so interesting.

---

[2] http://www.xtec.cat/~aparra1/astronom/craters/anecdotese.htm

PART I:

A HISTORY OF COMETS AND CATASTROPHES

1. Victor Clube and the Cosmic Turkey Shoot

Before I began the research on this book, a friend of mine (who is a climate scientist at a major U.S. research facility) made an interesting find: a letter addressed to the Chief, Physics and BMD Coordinator of the European Office of Aerospace Research and Development, dated 4 June 1996, entitled *'The Hazard to Civilization from Fireballs and Comets'*,[3] by S. V. M. Clube. For the uninitiated, Clube is an astrophysicist at the University of Oxford. In this short, four-page letter and summary statement, Clube writes (emphases in the original):

> (A) Asteroids *which pass close to the Earth* have been fully recognized by mankind for only about 20 years. Previously, the idea that substantial *unobserved objects* might be close enough to be a potential hazard to the Earth was treated with as much derision as the *unobserved aether*. Scientists of course are in business to establish broad principles (e.g., relativity) and the Earth's supposedly uneventful, uniformitarian environment was already very much in place. The result was that scientists who paid more than lip service to objects close enough to encounter the Earth did so in an atmosphere of barely disguised contempt. Even now, it is difficult for laymen to appreciate the enormity of the intellectual blow with which most of the Body Scientific has recently been struck and from which it is now seeking to recover.

What intellectual blow might Clube be talking about here? After a bit of thought, it occurred to me that he must be talking about the Comet Shoemaker-Levy fragment impacts on Jupiter which produced a huge amount of excitement at the time which was just two years before the date of this letter. To return to Clube's report, he continues:

> The present report, then, is concerned with those *other celestial bodies* recorded by mankind since the dawn of civilization which either miss or impinge upon the

---

[3] http://www.dtic.mil/cgi-bin/GetTRDoc?AD=ADA359871&Location=U2&doc=
GetTRDoc.pdf

Earth and which have also been despised. Now known respectively as comets (>1 kilometre in size) and meteoroids (<10 metres in size), it is the fireballs or "signs" produced by these meteoroids when they encounter the atmosphere which have generally been the greatest source of concern. Thus the fireball flux is noticeably augmented for a generation or more every other century or so, that is, whenever the Earth makes repeated passages through a fresh stream of cometary debris à la Shoemaker-Levy. Such augmentations have commonly been interpreted as evidence of impending global disaster or world-end due to the larger debris in train.

That he immediately switched from asteroids to comets seems to confirm my speculation that he was talking about Shoemaker-Levy. But it gets more interesting:

Confronted on many occasions in the past by the prospect of world-end, national elites have often found themselves having to suppress public panic – only to discover, too late, that the usual means of control commonly fail. Thus an institutionalized science is expected to withhold knowledge of the threat; a self-regulated press is expected to make light of any disaster; while an institutionalized religion is expected to oppose predestination and to secure such general belief in a fundamentally benevolent deity as can be mustered. [...]

(B) The present report based on the above grant addresses a variety of issues within the broad context of the hazard to civilization due to fireballs and comets. It consists of:

1. A brief statement of conclusions arising from a narrative report (3 copies);
2. A narrative report (with appendix) linking the results of three scientific studies described in papers submitted to mainstream journals (3 copies);
3. The relevant papers detailing the results which arise through the granted funds due to (a) Clube; (b) Clube & Napier; and (c) Clube, Hoyle, Napier & Wickramasinghe (3 copies); and
4. A co-authored foundation paper by Asher & Clube detailing the results from which items 3 and 2 progressed.

[...] It is emphasised here that the present report expresses a viewpoint which is contrary to the mainstream scientific theme currently reinforced through various US agencies in the wake of recent major findings under US leadership (e.g., those of Luis Alvarez, Eugene Shoemaker, David Morrison, etc). Despite the importance of this mainstream theme, it is recognized here that the cometary signatures in the terrestrial record are generally stronger than the asteroidal signatures in the case of both long term **and** short term effects, i.e., those affecting biological and geological evolution on the one hand **and** mankind and civilization on the other . . . .

There are fundamental paradoxes to be assimilated as a result of this unexpected situation. Thus the perceived culture of enterprise and enlightenment which underpins the two centuries culminating with the Space Age and which led mankind to spurn comets and fireballs may now be seen as the prelude to a profound paradigm shift: the restoration of an environmental outlook more in keeping with that which preceded American Independence and which paid serious heed to comets and fireballs. [...]

Clube then thanks the USAF for *"its generous and timely injection of funds"* and we note that the letter was cc'd to, among others, Edward Teller at the Hoover Institute, S. Fred Singer at Fairfax, VA, and Jack A. Goldstone, Davis, CA.

I should add that I obtained the whole report and the papers it refers to (and more besides).

The report makes one startling remark that I'm going to include because it is extremely interesting (again, with Clube's emphasis):

[T]he Christian, Islamic and Judaic cultures have all moved since the European Renaissance to adopt an unreasoning anti-apocalyptic stance, apparently unaware of the burgeoning science of catastrophes. History, it now seems, is repeating itself: it has taken the Space Age to revive the Platonist voice of reason but it emerges this time within a modern anti-fundamentalist, anti-apocalyptic tradition over which governments may, as before, be unable to exercise control. ... **Cynics (or modern sophists), in other words, would say that we do not need the celestial threat to disguise Cold War intentions; rather we need the Cold War to disguise celestial intentions!**

Just think about what this might mean, considering when it was written and all that has happened since.

The summary of conclusions accompanying this letter is equally interesting, and we'll discuss it shortly. But first, a little bit of background.

I often get accused of 'fear mongering' because again and again in my writing I bring up the subject of swarms of comets and comet fragments that have repeatedly barreled through the solar system, wreaking havoc and bringing death and destruction to earth. I even think that it is fascinating that the big breakthrough in my experiment in Superluminal Communication [4] came on the day that the fragments of Comet Shoemaker-Levy began striking Jupiter – even at the very moment of the first impact – and that this communication with 'myself in the future' has focused so much attention on the topic. As a

---

[4] http://www.cassiopaea.org

result of the research prompted by this communication, I wrote an entire 800-page book, *The Secret History of The World*, that is woven around the issue of cometary explosion-type catastrophes that obviously have occurred repeatedly throughout history.

In the early days of publishing the results of this experiment, I was nonplussed by the many attacks I came under from all quarters. I was accused of 'channeling aliens' (not true), of wanting to 'start a cult' (what is cultic about doing research into scientific subjects and exposing religion for the fraud it is?) and so on. That sort of thing really hurt and puzzled me at first, but I have now seen it for the blessing it was: it has helped me to learn about the kinds of people who are in charge of our world, the kind of people who want to keep secrets so that they can hang onto their power; the kind of people who create such things as the 'War on Terror' to conceal from the masses of humanity the future that may very well bring our civilization to an end; the kind of people who know that survival of cometary bombardment is possible and who want to be the only ones who do survive, and to hell with everyone else.

Mike Baillie, in his book about the Black Death, discussed in more depth later, writes (emphasis mine):

> It is increasingly evident that intellectually the world is divided into two. There are those who study the past, in the fields of history and archaeology, and see no evidence for any human populations ever having been affected by impacts from space. In diametric opposition to this stance there are those who study the objects that come close to, and sometimes collide with, this planet. Some serious members of this latter group have no doubt whatsoever that there must have been numerous devastating impacts in the last five millennia; the period of human civilization. In a paper published in 2005, David Asher and colleagues have looked at the objects that are known to have come close to the earth in recent times. *They conclude, based on various strands of evidence (for example, the number of meteorites discovered on earth that originated on the moon) that the average time between impacts on earth is no more than 300 years, probably less.* (Baillie 2006, 199)

Checking the authors Baillie is referring to, we find Bill Napier listed, a colleague of Victor Clube. This brings us to another division. There is a longstanding debate going on about this issue, as was mentioned by Clube in the first parts of the letter quoted above. He wrote:

> It is emphasised here that the present report expresses a viewpoint which is contrary to the mainstream scientific theme currently reinforced through various US agencies in the wake of recent major findings under US leadership (e.g., those of Luis Alvarez, Eugene Shoemaker, David Morrison, etc). Despite the importance

of this mainstream theme, it is recognized here that the cometary signatures in the terrestrial record are generally stronger than the asteroidal signatures in the case of both long term **and** short term effects, i.e., those affecting biological and geological evolution on the one hand **and** mankind and civilization on the other. The raison d'etre behind this situation is a cometary input dominated in the long term by objects >100 kilometres in size which substantially break up in the short term into objects <1 kilometre in size, the "window" of significance so far as the average interval between random impacts by comets and asteroids in the intervening size range are concerned being approximately 1-10 million years. To concentrate, for planetary defence purposes, on catastrophes which occur only within this particularly narrow range of frequencies is patently absurd.

Clube's reference to the 'mainstream scientific' ideas about comets and asteroids and so on is only the tip of the iceberg in reference to this debate.

The debate is about asteroids vs. comets. Asteroids are solid bodies of rock and there are about 1,000 of them with diameters of 1km or more that cross the orbit of the earth. They are called 'Apollo' or 'earth-crossing' asteroids. The 'American school' of astronomers believe that these objects are the main threat to earth and humanity and they are concerned with finding them, tracking them, and working out their orbits. This school believes that if all these asteroids can be mapped, and any 'bad ones' dealt with, earth will be

Apollo Asteroid MD 2011 as seen by the Faulkes Telescope South in a single 20-second exposure taken with a red-green-blue filter.

safe for the foreseeable future. Their estimates are that we only get hit with one of these 'earth-crossing' asteroids every 100,000 years or so.

At this point in time, the American school of astronomers has already found and tracked about 700 of the estimated 1,000 such asteroids and, so far, none of them are likely to hit the earth anytime soon. By the end of 2008, they expected to have located 90% of these potential threats. Of course, they aren't talking about objects smaller than 1 km because they are believed to pose much less risk even if they do smack into the earth. So it is that the American school believes that they can, over time and with superior American technology, survey everything around us and keep our place in space 'under control'.

What they are saying, as Baillie astutely points out, is this: There are objects that cross the path of the earth that hardly ever hit us (only about every 100,000 years), but they are going to make us safe by finding any and all of them and devising methods to take out the ones that *might* pose a threat at any point in the future. They assume, of course, that if they figure out that any of them might be a threat by mapping their orbits, they will have time to do this. Near Earth Asteroid '2012 DA14' is a case in point. This 44 meter chunk of rock that was discovered in 2012 is projected to pass our planet at a mere 16,800 miles on February 15th 2013. NASA have suggested that the asteroid could be 'painted' in an effort to affect its ability to reflect sunlight, thereby changing its temperature and altering its spin and, theoretically, moving it off course. The problem however, is that it would take 2 years to build a spacecraft to carry out the job.[5]

It's obvious that Victor Clube is not a member of the American school.

The 'Comet Hazard' school, on the other hand, is British-based and they think very differently from the American 'mainstream' asteroid school.

Comets are said to be different from asteroids because they are made up of water ice, frozen gas, organic materials, and odd bits of rock and metal. The standard theory (which may need revision according to those who advocate the electric universe theory) says that comets are heated as they pass through the solar system and this causes outgassing. It is then that we see them as bright objects with long tails.

After a few circuits through the solar system, some comets outgas completely and all that is left is a very black lump of any size, typically at least a few kilometers in diameter. The reason a worn-out comet is so black is possibly due to the polyaromatic hydrocarbons that are concentrated onto the comet's surface like a coating of tar. Such objects, unlike asteroids, are very difficult to spot because they do not reflect light.

---

[5] http://www.rt.com/news/paint-asteroid-earth-nasa-767/

Comets also leave trails of dust and debris in the inner solar system and the earth passes through such periodically. When this happens, there are generally meteor showers, which are really particles of comets burning up in the atmosphere.

Comets can also break up into smaller – but still sizable – chunks.

Now, imagine that in a trail of comet dust, there are also some fairly large chunks of black, unseeable comet fragments. If you can't see them, you can't do anything about them. And when they do 'hit', they tend to burn up and/or explode violently in the atmosphere (e.g., the Tunguska event). So they don't leave long-lasting traces such as craters for archaeologists to find and say, 'Yes, the fall of this civilization was due to an assault from outer space.' No, there is only fire, death and destruction – sometimes total.

What all this means is that the comet problem does not submit itself to an efficient solution.

The Comet Hazard school scientists propose that the Tunguska event was due to a fragment of Comet Encke. These scientists also now have the *fact* of the fragments of Comet Shoemaker-Levy hitting Jupiter in July of 1994 to illustrate the problem we face.

The Comet Hazard scientists also think, as mentioned above, that impacts are a lot more frequent than many people suppose.

So, to sum it up: there are two very different schools that study hazards from space. The asteroid school says that there have been very few impacts and the problem is solvable, and the comet school says there is evidence that there have been numerous impacts by comet debris that have had profound effects on human civilizations and will again, probably very soon.

With that said, let's take a look at Victor Clube's 'Summary statement of conclusions' about fireballs and meteorites that he attached to his cover letter to the *European Office of Aerospace Research and Development* just two years after Shoemaker-Levy hit Jupiter and five years before 11 September 2001. He writes:

Asteroid strikes, though important, are not the most serious short-term risk to mankind or civilization.

Every 5-10 generations or so, for about a generation, mankind is subject to an increased risk of global insult through another kind of cosmic agency.

Every 5 to 10 generations? That's a pretty shocking statement. If it is true, then why don't we know about this? Why don't historians know about it? Why don't average people who learn history in school know about these things?

This cosmic agency is a "Shoemaker-Levy type" train of cometary debris resulting in sequences of terrestrial encounters with sub-km meteoroids.

While the resulting risk is ~10%, the global insults take the form of (a) multiple multi-megaton bombardment, (b) climatic deterioration through stratospheric dust-loading, not excluding ice-age, and (c) consequent uncontrolled disease/plague.

The sequence of events affecting involved generations is potentially debilitating because, *whether or not the risk is realised*, civilization commonly undergoes violent transitions, e.g., revolution, migration and collapse.

In short, whether or not there are any impacts during those periods when 'something is out there, rather close and threatening', people get a little upset when they get the feeling that they are living on a target in a cosmic shooting gallery. The knowledge that the earth beneath our feet may not be so firmly and peacefully fixed in space naturally assaults our deepest feelings of security. It's almost as if Clube is saying that there is some sort of contagious madness, a stampeding of human beings, almost, like a herd of cattle stampeding over a cliff because someone accidentally (or purposely) shoots a gun into the air. That's not even a bad metaphor because, as we are going to see, it seems that the ruling elite *do* tend to take advantage of such conditions for their own purposes, which are usually to grab more power and plunder.

Subsequently perceived as pointless, such transitions are commonly an embarrassment to national elites even to the extent that historical and astronomical evidence of the risk are abominated and suppressed.

Indeed, when the madness dies down and the people begin to realize what fools they have made of themselves and, more importantly, what fools their leaders are; when they view how much death and destruction has occurred for no good reason at all except a form of madness, I'm sure that the elites do want to just shove it all under the rug and try to make everyone forget that it ever happened so as to keep their hands on the reins of power.

As we will see, this isn't how it always turns out.

Sometimes, the people are so hostile when they see how they have been abused by their leaders, the leaders pay a rather high price ... sometimes their very heads.

Upon revival of the risk, however, such "enlightenment" becomes an inducement to violent transition since historical and astronomical evidence are then in demand.

Such change and change about in addition to the insult is evidently self-defeating and calls for a procedure to eliminate the risk.

The term 'enlightenment', used above, is a reference to people waking up to what is possibly going on out there in space.

Our technological ability to counter (a) multiple multi-megaton bombardment and (b) stratospheric dust-loading should therefore be explored.

The very short lead-time commonly associated with the detection of sub-km meteoroids approaching the Earth implies countering procedures which differ from those associated with catalogued km-plus asteroids and comets.

So, the question is: if there is even a 10% chance that we are facing a Shoemaker-Levy-type event, why isn't anybody doing anything about it?

Well ... maybe they are. Maybe all this 'War on Terror' business and getting control of resources is, at its root, the psychopaths' way of handling a threat to their survival. Since it is obvious that the 'War on Terror' was created as a direct replacement for the 'Cold War', maybe it is true that "we do not need the celestial threat to disguise 'War on Terror' intentions; rather we need the 'War on Terror' to disguise celestial intentions!" Maybe it isn't the 'Twilight of the Psychopaths' as Dr. Kevin Barrett might like to think, but the Twilight of Humanity; if we don't wake up.

Anyway, I dug around a bit, following references from Clube, and found that there is, in fact, a group that is looking at these things, but I don't think they are doing it to inform the general public, nor do I think that they have the best interests of the public in mind. Have a look at the INSAP website[6] and follow some of their links. Their first conference, attended by Clube and referenced obliquely in his report on the 'Hazards to Civilization', was held at the Mondo Migliore, under the sponsorship of the Vatican Observatory, Rocco di Papa, Italy, from 27 June–2 July 1994. Their mandate reads:

> INSAP conferences explore the rich and diverse ways in which people of the past and present incorporate astronomical events into literary, visual, and performance arts.
>
> This emphasis distinguishes INSAP from other conferences that focus on archeoastronomy, ethnoastronomy, or cultural astronomy.
>
> INSAP provides a mechanism for a broad sampling of artists, writers, musicians, historians, philosophers, scientists, and others to talk about the diversity of astronomical inspiration.

This, of course, reminds me of the strange recent news item about the new Pope evicting the Jesuit astronomers from the papal summer palace.[7] Fol-

---

[6] http://www.insap.org/insap/

[7] https://www.sott.net/articles/show/146673-Pope-tells-astronomers-to-pack-up-their-telescopes

lowing that story, one then finds this: 'Italian scientists attack Pope's equivocation on Galileo trial':

> Pope Benedict XVI has been forced to cancel a visit to the prestigious La Sapienza University in Rome after lecturers and students expressed outrage at his past defence of the Catholic church's actions against Galileo.
>
> The Pope had been due to make a speech at the university on Thursday 17 January 2008. …
>
> Sixty-seven academics have said that the Pope effectively condoned the 1633 trial and conviction of the astronomer Galileo for heresy, in remarks he made while head of the Sacred Congregation for the Doctrine of the Faith, the successor to the notorious Inquisition.
>
> As Cardinal Josef Ratzinger, Pope Benedict said that Galileo had turned out to be correct about the earth revolving around the sun, and that subsequent biblical scholarship had rejected literalist readings of texts that had been taken by the Church to deny this. Nevertheless, he said, Galileo had been dogmatic and sectarian in his statements at the time, and the Church authorities had acted reasonably given the levels of knowledge available then.
>
> But the scientists say that this is "insulting" and unacceptable equivocation. The Church was unjust, irrational and unfair in its treatment of their predecessor and its outright rejection of Copernican theory, they say.[8]

Apparently, before Ratzinger was selected to run the Catholic Fraud Factory, the Jesuits were pretty interested in figuring out what was going on on our planet – for what purposes, we may never know.

Clube was there at one of INSAP's meetings and presented a paper titled: *'The Nature of Punctuational Crises and the Spenglerian Model of Civilization'*.[9] Parts of it are a bit difficult to read, but it is well worth the trouble of reading all the way through – maybe more than once – and giving a lot of thought to the implications of what he writes there, especially in regard to any group of people who are trying to dig out this kind of information and present it to the public. Clube makes it abundantly clear why this must be considered a revolutionary activity.

What strikes me as particularly obtuse is the way the American school of asteroid impact is going about maintaining the party line. Apparently, under the influence of the British school of cometary bombardment, they are think-

---

[8] http://www.ekklesia.co.uk/node/6595

[9] https://www.sott.net/articles/show/147270-The-Nature-of-Punctuational-Crises-and-the-Spenglerian-Model-of-Civilization

ing about all of these issues. And from the recent AIAA 2007 Planetary Defense Conference we note what is agitating them most:

> An asteroid impact could occur anywhere on the globe at any time, so planetary defense has implications for all humankind. All nations on Earth should be prepared for this potential calamity and work together to prevent or contain the damage. That said, there is currently very little discussion or coordination of efforts at national or international levels. No single agency in any country has responsibility for moving forward on NEO deflection, and disaster control agencies have not simulated this type of disaster.
>
> Providing funding over the long term was also seen as a challenge. Much of the work in virtually all areas of planetary defense has been done on individuals' own time and initiative. There is a need for ongoing studies and peer-reviewed papers to improve our knowledge in this area, as well as to increase the credibility of the issue and the public's trust in our ability to respond. The reality is that NEO deflection or disaster mitigation efforts may not be required for decades or longer, so governments, which are focused on more immediate concerns, may not be willing to commit sufficient recourses to this type of work. Determining the appropriate level of this work and funding such activities over the long term is seen as a major issue.
>
> In addition, major legal and policy issues related to planetary defense need to be resolved. An example is *liability for predictions that prove false or deflection missions that only partially work or fail completely*, resulting in an impact. Other examples include:
> - *A prediction is made that an impact may occur in a specific area, and residents and businesses that might be affected leave. Are there liabilities associated with the loss in property values if the prediction is wrong?*
> - *A nation makes a deflection attempt, but it fails to change the object's orbit enough to miss Earth. Is that nation now responsible for the damage inflicted?*
> - A NEO threat demands the nuclear option, but public perception is that the possibility of a launch failure and subsequent damage is more acute than the threat from the NEO. What are the liabilities and political and policy implications associated with a launch failure during a deflection mission?
>
> These types of issues should be discussed and resolved before they are raised by a serious threat.[10]

One would have thought that legal liability would be the least of mankind's worries in the event of an asteroid impact.

---

[10] http://www.sott.net/articles/show/147239-AIAA-2007-Planetary-Defense-Conference

Turning back to Clube's full-text report, on page 2, discussing potential impacting giant comet remnants, we read that:

> [T]heir presence is readily enough betrayed by the zodiacal dust which continues to accumulate in the ecliptic and by the rather sudden encounters which the Earth makes **every other century or so**, for several decades ... These encounters produce an overabundance of fireballs, penetrating the Earth's atmosphere, implying both an increased probability of bombardment by sub-kilometer debris and an increased risk that the Earth will penetrate the core of a minor disintegration stream à la Shoemaker-Levy.

If this is the case, what is the documentation for these events that Clube suggests are routinely covered up by embarrassed 'governing elites'? Not only is nobody doing anything about the threat, historians seem to ignore the issue completely. That is one of the things that makes this research so difficult. It's not just a matter of picking up a history book and the author saying something like: 'In 325 AD Constantine was terrified by an overhead cometary explosion and decided to adopt Christianity as a consequence, and to make it the state religion.' As we'll see further on however, this is actually pretty close to the truth. I'm reminded of the historian Herodian's report of the siege of Aquileia by Maximinus in the 230s during which operation the soldiers saw 'the god Apollo' appearing 'frequently' above the city and fighting for it. Herodian wasn't certain whether the soldiers *really* saw it or whether they just invented it to explain their defeat. The standard explanation is, of course, that it was common for generals to claim 'appearances' in order to give heart to their troops. But maybe, sometimes, they *did* see something? (This reminds me of something else: I recently read a news article about a fellow who had a meteorite come through the roof of his house while he was at work. His reaction was extremely interesting: he announced that this was a 'sign from God' that he needed to go to church and renew his faith.)

As it happens, further on in the narrative we find out just what periods Clube is referring to:

> There have been five extended epochs since the Renaissance when the Earth apparently encountered the fragmentation debris of previously unsighted comets.

Well, we know from the work of Mike Baillie that the period around 540 AD is highly suspect, as is the period around the Black Death. (See his books *From Exodus to Arthur* and *New Light on the Black Death*.) The events that Baillie suggests occurred during those periods are backed up by very strong scientific data. But those aren't the periods that Clube is talking about here.

He is saying *"since the Renaissance"*. The Renaissance, of course, followed closely on the heels of the Black Death, which Baillie considers to have been a period of cometary bombardment that killed almost half of humanity. (Or so it seems from the statistics given for those areas where statistics were obtainable.) In the broadest of terms, the Renaissance covers the 200 years between 1400 and 1600, although specialists disagree on exact dates. The Hundred Years War covers the 116-year period from 1337 to 1453 and the Black Death began in 1347/1348, so it could even be inferred that the Black Death was the gestational period for the Renaissance, or that the Renaissance occurred as a reaction to the Black Death.

What we now see is that Victor Clube is suggesting that there was a lot more going on in our recorded history than we presently know about, and that the rise and fall of nations and civilizations may be closely related to what is going on out there in space. To continue:

> During these epochs, broadly coinciding with the Hundred Years' War, the Reformation, the Thirty Years' War (including the English Civil War), the French Revolutionary Period (including the American War of Independence) and the mid-nineteenth century Revolutionary crisis in Europe [including the American Civil War], the various national authorities could do very little to restrain public anxiety in the face of the perceived danger.

We now have some specific periods where Clube *et al.* believe that strange things were going on in the space around our planet. It might help us to better understand our own time period to take a look at times past, starting with the Black Death.

2. Mike Baillie
and the Cosmic Pestilència

Dr. Mike Baillie is a Professor Emeritus of Palaeoecology in the School of Archaeology and Palaeoecology at Queen's University Belfast, Northern Ireland. He is a leading expert in dendrochronology, or dating by means of tree-rings. In the 1980s, he was instrumental in building a year-by-year chronology of tree-ring growth reaching 7,400 years into the past. His book *From Exodus to Arthur: Catastrophic Encounters with Comets*, published in 1999, is a must-read. But his most recent book, which I mentioned in the last chapter, is what I want to write about.

New Light on the Black Death, published by Tempus in 2006, is an intense book. Baillie, a real scientist who cannot be easily maligned as a crackpot, is saying that the Black Death, one of the most deadly pandemics in human history, said to have killed possibly two thirds of the entire population of Europe, not to mention millions all over the planet, probably wasn't bubonic plague but was rather Death By Comet(s).

Baillie has the scientific evidence to back up his theory, and his evidence actually supports – and is supported by – what the people of the time were saying: earthquakes, comets, rains of death and fire, corrupted atmosphere, and death on a scale that is almost unimaginable. Most people nowadays are not really aware of what happened just 665 years ago. (The inquiring mind immediately wonders what might happen when we hit 666 years after. That would be 2013 …)

It is worth pausing a moment to consider the numbers. China, where the Black Death is said to have originated, lost around half of its entire population (declining from around 123 million to around 65 million). Recent research into European death tolls also suggests a figure of 45% to 50% of the total European population dying during a four-year period though the figure fluctuated from place to place. In Mediterranean Europe – Italy, the South of France and Spain – where the plague ran for about four years consecutively, it was probably closer to 70–75% of the total population. (In the US today that would be equivalent to reducing the population from its current 305 million to 75 million in less than four years. That would also amount to having to bury or dispose of around 225 million corpses.) In Germany and England it was probably closer to 20%. Northeastern Germany, Bohemia, Poland and Hungary are believed to have suffered less for some reason (and there are a few theories which are not entirely satisfactory).

There are no estimates available for Russia or the Balkans, so it seems that they may have suffered little, if at all. Africa lost approximately 1/8th of its population (from around 80 million to 70 million). All of these figures actually highlight one of the problems that Baillie brings up: the variability of death rates according to location. Whatever the death rate in any given location, the bottom line is that the Black Death produced the largest death toll from any known pandemic in recorded history and, as Baillie points out, nobody really knows what it was. Certainly, the identification of the pandemic as bubonic plague has been accepted as 'known' for a very long time, so how is it that Baillie questions this well-established fact? It turns out he's not the only one.

In 1984, Graham Twigg published *The Black Death: A Biological Reappraisal*, where he argued that the climate and ecology of Europe and particularly England made it nearly impossible for rats and fleas to have transmitted bubonic plague and that it would have been nearly impossible for *Yersinia pestis* to have been the causative agent of the plague, let alone its explosive spread across Europe during the 14th century. Twigg also demolished the common theory of entirely pneumonic spread. He proposed, based on his examination of the ev-

An illustration of the Black Death from the Toggenburg Bible.

idence and symptoms, that the Black Death may actually have been an epidemic of pulmonary anthrax caused by *Bacillus anthracis*.

Another researcher who is skeptical of the standard model is Gunnar Karlsson who, in 2000, pointed out that the Black Death killed between half and two-thirds of the population of Iceland, although there were no rats in Iceland at this time. (See Karlsson's *The History of Iceland*.) Baillie sums up the problem as follows:

> The Black Death of 1347 was believed to be the third great outbreak of bubonic plague; a plague that is traditionally spread by rats and fleas. The previous instances were the Plague of Athens in 430 BC and the plague at the time of Justinian which arrived into Constantinople in AD 542. The Plague of Athens was described by Thucydides, while the Justinian plague was described by Procopius, among others. ...
>
> The plague is supposed to have originated in Central Asia, or somewhere in Africa, where plague is endemic in some rodent populations. It is assumed that some environmental stimulus caused infected rodents to leave their normal habitats and infect rat populations, and ultimately human populations, in areas where there was no natural immunity. The mechanism of transfer is believed to have been infected fleas leaving the bodies of dead rats and moving to human hosts who were in turn infected by the feeding fleas. It is believed that trade routes brought the disease to the Black Sea region and from there to the central Mediterranean by late 1347. It was then introduced into Europe through northern Italy and southern France. It immediately started killing people in large numbers spreading overland at about 1.5 km per day. Between January and the summer to autumn of 1348 it had spread as far as the British Isles, and by 1350 to Scandinavia and eventually even Iceland. The spread seems to have curled up through France, across Belgium into Germany and on into central southern Europe. This first wave burned itself out by 1351, though there was a second wave in 1361.
>
> It is generally believed that the plague hit an already weakened population in Europe. ...
>
> ... At its most basic, the problem is with those rats and fleas. For the conventional wisdom to work there have to be hosts of infected rats and they have to be moving at alarming speed – you would almost have to imagine infected rats scuttling ever onward (mostly northward) delivering, as they died, loads of infected fleas. The snags with this scenario are legion. For example, there are no descriptions of dead rats lying everywhere (this is explained by suggesting that either the rats were indoors, or people were so used to dead rats that they were not worth mentioning; though if they were indoors how did they travel so fast?). It did not seem to matter whether you were a rural shepherd or cleric or a town dweller, both were infected. Yet strangely with this very infectious disease some

cities across Europe were spared. Moreover, these rats must have been happy to move to cool northern areas even though bubonic plague is a disease that requires relatively warm temperatures. Then, when there are water barriers, these rats board ships to keep the momentum going. (Baillie 2006, 11–12)

O. J. Benedictow, an advocate of the rats-and-fleas scenarios, and quoted by Baillie, tells us about these amazing creatures in his 2004 book *The Black Death 1346-1353: The Complete History*:

> The Black Death's strategic genius made also another masterstroke that greatly increased the pace of its conquest of the Iberian peninsula. Shortly after its multiple invasions of important urban centres along the coast of the Kingdom of Aragon, it performed a remarkable metastatic leap and arrived triumphantly in the town of Santiago de Compostela in the very opposite, north-westernmost corner of the Iberian Peninsula. (Benedictow, quoted in Baillie 2006, 14)

In 2001, epidemiologists Susan Scott and Christopher Duncan from Liverpool University proposed the theory that the Black Death might have been caused by an Ebola-like virus, not a bacterium. Their research and findings are thoroughly documented in *Biology of Plagues*. More recently the researchers have published computer modeling demonstrating how the Black Death has made around 10% of Europeans resistant to HIV. (See *Return of the Black Death: The World's Greatest Serial Killer* and *Biology of Plagues: Evidence from Historical Populations*, both by Susan Scott and Christopher J. Duncan.)

In a similar vein, historian Norman F. Cantor, in his 2001 book *In the Wake of the Plague*, suggests the Black Death might have been a combination of pandemics including a form of anthrax and a cattle murrain. He cites many forms of evidence including: reported disease symptoms not in keeping with the known effects of either bubonic or pneumonic plague, the discovery of anthrax spores in a plague pit in Scotland, and the fact that meat from infected cattle was known to have been sold in many rural English areas prior to the onset of the plague.

Samuel K. Cohn, quoted extensively by Baillie, also rebutted the theory (and that's really all it is, and a weak theory at that) that the Black Death was bubonic plague. In the *Encyclopedia of Population*, he points to five major weaknesses in this theory:

- *very different transmission speeds:* The Black Death was reported to have spread 385 km in 91 days in 664, compared to 12-15 km a year for the modern bubonic plague, which has the assistance of trains and cars.
- *difficulties with the attempt to explain the rapid spread of the Black Death*

by arguing that it was spread by the rare pneumonic form of the disease: in fact this form killed less than 0.3% of the infected population in its worst outbreak in Manchuria in 1911.

• *different seasonality:* the modern plague can only be sustained at temperatures between 50 and 78°F (10 and 26°C) and requires high humidity, while the Black Death occurred even in Norway in the middle of the winter and in the Mediterranean in the middle of hot, dry summers.

• *very different death rates:* in several places (including Florence in 1348) over 75% of the population appears to have died; in contrast the highest mortality for the modern bubonic plague was 3% in Mumbai in 1903.

• *the cycles and trends of infection were very different between the diseases:* Humans did not develop resistance to the modern disease, but resistance to the Black Death rose sharply, so that eventually it became mainly a childhood disease.

Cohn also points out that while the identification of the disease as having buboes relies on the account of Boccaccio and others, they described buboes, abscesses, rashes and carbuncles occurring all over the body, while the modern disease rarely has more than one bubo, most commonly in the groin, and is not characterized by abscesses, rashes and carbuncles, which is what Boccaccio described.

The gist of Cohn's argument is that whatever caused the Black Death, it was not bubonic plague. (See also: Samuel K. Cohn [2002], 'The Black Death: End of the Paradigm', *The American Historical Review* 107: 703–738 and *The Black Death and the Transformation of the West* [European History Series] by David Herlihy and Samuel K. Cohn.)

When one begins to dig into the subject, we find that there was one study claiming that tooth pulp tissue from a fourteenth-century plague cemetery in Montpellier tested positive for molecules associated with *Y. pestis* (bubonic plague). Similar findings were reported in a 2007 study, but other studies have not supported these results. In fact, in September of 2003, a team of researchers from Oxford University tested 121 teeth from sixty-six skeletons found in fourteenth-century mass graves. The remains showed no genetic trace of *Y. pestis*, and the researchers suspect that the Montpellier study was flawed.

What these studies do not address is the problem that the apparent means of infection or transmission varied widely, from human-to-human contact as in Iceland (rare for plague and cutaneous *Bacillus anthracis*) to infection in the absence of living or recently dead humans, as in Sicily (which speaks against most viruses).

To all the problems with the bubonic plague theory cited above, we have to add what the contemporary writers recorded. Philip Ziegler collected many of these accounts in his book *The Black Death*, though he dismissed them as 'metaphor'. We'll be looking at some of them in just a moment.

Mike Baillie admittedly didn't start out to write a book about cometary impacts being implicated in the great pandemics of the past; he had just noticed some strange tree-ring patterns that happened to coincide with this historical catastrophe and thought that, perhaps, there was some sort of environmental downturn that weakened the human population, making humanity susceptible to bacterial or viral death on a large scale. But, what he found was a dangling thread that, once he began to pull on it, unraveled the 'accepted wisdom' about the Black Death and sent him off on a search that led to astonishing conclusions.

As mentioned, the first clue was tree rings – that's natural since Baillie is a dendrochronologist. He compared these tree rings to dated ice-core samples that had been analyzed and discovered a very strange thing: ammonium. There are, as it happens, four occasions in the last 1,500 years where scientists can confidently link dated layers of ammonium in Greenland ice to high-energy atmospheric interactions with objects coming from space: 539, 626, 1014, and 1908 (the Tunguska event). In short, there is a connection between ammonium in the ice cores and extra-terrestrial bombardment of the surface of the earth.

Notice the above statement that there are four events that can be definitively linked with high-energy interactions. Baillie presents the research in his book showing that the exact same signature is present at the time of the Black Death in both the tree rings and in the ice cores, *and* at other times of so-called 'plague and pandemic'.

As it happens, the ammonium signal in the ice-cores is directly connected to an earthquake that occurred on 25 January 1348 – and Baillie discovers that there was a fourteenth-century writer who wrote that the plague was a *"corruption of the atmosphere"* that came from this earthquake.

How could a plague come from an earthquake, one may ask?

Baillie points out that we don't always know if earthquakes are caused by tectonic movements; they could be caused by cometary explosions in the atmosphere or even impacts on the surface of the earth.

In *Rain of Iron and Ice*, John Lewis, Professor of Planetary Sciences at the Lunar and Planetary Laboratory, co-director of the NASA/University of Arizona Space Engineering Research Center, and Commissioner of the Arizona State Space Commission, tells us that the earth is regularly hit by extraterrestrial objects and many of the impacting bodies explode in the atmosphere, as happened in Tunguska, leaving no craters or long-lasting visible evidence of a body from space.

But just because there is no long-lasting evidence doesn't mean there is no significant effect on the planet and/or its inhabitants. These impacts or atmospheric explosions may produce earthquakes or tsunamis without any witnesses being aware of the cause. After all, the earth is 75% water and any eyewitness to such an event would probably never live to tell about it, so we really have no way of knowing if all the earthquakes on our planet are tectonic in nature or not.

Lewis points out:

> In an average year there is one atmospheric explosion with a yield of 100 kilotons or more. The large majority occur in such remote areas, or so high in the atmosphere, that they are not observed. Even if observed, the witnesses may see only a flash of light in the distance, or hear the 'rumble of distant thunder' coming from the open oceans. Thus even those that are observed are often not recognized. (Lewis, quoted in Baillie 2006, 128)

As Baillie points out, Lewis is talking about a 'typical' year and it is obvious from other studies that not all years are equal – some are less typical than others. Baillie writes:

> As Lewis pointed out, we know from many strands of evidence broadly what the impact rate should be over time. The fact that impacts are not in the historical record [or not admitted or discussed by historians or archaeologists] is not because none happened. After all, there are those well-attested crater fields that were formed in the last few millennia in Estonia, Poland, Germany and Italy – which were not recorded historically; their existence was deduced from holes in the ground. So we know the recording mechanism is flawed!
>
> What needs to be added ... is one key piece of intuitive thinking. Here is a quote from [Lewis'] Scenario D:
>
>> (In this scenario) In 1946 a 25,000-metric-ton achondritic fireball explodes at 4:00 a.m. local time at a height of 11km above Fergana, Uzbekistan. The 1-megaton blast damages buildings over an area several kilometers in diameter, searing the area with intense heat and setting thousands of fires. The fires burn out of control, killing 4,146. Over 20,000 residents are awakened by the brilliant flash of light and heat to find their city in flames. An 'earthquake' is reported by the survivors. Several metric tons of meteorite fragments are mixed in with the debris of 2000 burned-out and collapsed buildings, where they are indistinguishable from scorched and blackened fragments of structural brick and rock. (Baillie 2006, 129)

The point of this is that there is almost no way to monitor whether or not any given disaster or catastrophe is definitively an impact as opposed to a violent earthquake. The result is that centuries could be passing, with numerous cometary impacts happening all the time, and no one suspecting the true hazards from space. As Baillie points out: there are many earthquakes recorded in history, but *no* impacts. And yet, there is the evidence that the impacts *have* happened – on the ground, and in the ice cores. And there is Tunguska.

Reports of the Tunguska event tell us that the ground shook around the impact/explosion zone for a radius of about 900 km. At the time of any larger impact event, the earthquake would be proportionally more severe. Any individuals far enough away to survive such an event, would only have seen a flash, felt a tremor, and heard a loud rumbling noise. If they were too far away to see the flash, or were indoors, they would only report an earthquake.

In short, what the work of Lewis brings to the table is the idea that some well-known historical earthquakes could very well have been impact events. Baillie mentions that one obvious prospect is the great Antioch earthquake of AD 526 which was described by John Malalas (E. Jeffreys, M. Jeffreys and R. Scott, *The Chronicle of John Malalas* [Byzantina Australiensia 4], Melbourne: Australian Assoc. Byzantine Studies 4, 1986):

> … those caught in the earth beneath the buildings were incinerated and sparks of fire appeared out of the air and burned everyone they struck like lightning. The surface of the earth boiled and foundations of buildings were struck by thunderbolts thrown up by the earthquakes and were burned to ashes by fire … it was a tremendous and incredible marvel with fire belching out rain, rain falling from tremendous furnaces, flames dissolving into showers … as a result Antioch became desolate … in this terror up to 250,000 people perished. (Malalas, quoted in Baillie 2006, 130)

Baillie also points out that a series of such impacts and overhead explosions, would more adequately explain the longstanding problem of the end of the Bronze Age in the Eastern Mediterranean in the 12th century BC. At that time, many – uncountable – major sites were destroyed and totally burned and it has all been blamed on those semi-supernatural 'Sea Peoples'. If that was the case, if it was invasion and conquest, there ought to at least be some evidence for such a scenario, like dead warriors or signs of warfare, but for the most part, that is not the case. There were almost no bodies found and no precious objects except those that were hidden away as though someone expected to return for them, or didn't have time to retrieve them. The people who fled (extra-terrestrial events often have precursor activities and warnings because a comet can be observed approaching for some time) were probably also killed in the very act of fleeing and the result was total abandonment and total destruction of the cities in question.

And the onset of Dark Ages.

So, the possibility that many destructions of the past could have been related to impact events has never been taken seriously or tested and this could be a perilous error.

The question that Baillie asks, but never really answers, is: What was it that so successfully stopped people asking why there is a traditional and deeply ingrained fear of comets in the psyche of humanity? He points out that, yes, there are people outside of mainstream academia who ask these questions. But why, against all good common sense, is this subject so widely and systematically ignored, marginalized and ridiculed?

The odd thing is that, even though Baillie points out that many high-level scientists and government agencies are taking these things seriously (Lewis, for example), it is still ignored, marginalized and ridiculed to the general public via the mainstream media. Baillie writes:

> In case readers think this is simply rhetoric, this is as good a place as any to mention a forthcoming event. On 13 April 2029 an asteroid named Apophis will pass by the earth at a distance of less than 50,000 km. If you're alive at the time, and it is not cloudy, you'll be able to see it pass with the naked eye. Apophis is more than 300 m in diameter. If, as it passed the earth, it just happens to pass through a certain narrow window in space, then, in 2036 it will return and hit the earth (this narrow window is a point where the earth's gravity would deflect the orbit of Apophis just enough to ensure an impact in 2036). If Apophis hits the earth the impact will be in the 3000-megaton class. It is entirely reasonable to state that such an impact, taking place anywhere on the planet, would collapse our current civilization and return the survivors, metaphorically speaking, to the Dark Ages (it is believed that in such an event globalised institutions, such as the financial and insurance markets would collapse, bringing down the entire interconnected monetary, trade and transport systems). Impacts from space are not fiction, and it seems highly likely that quite a number have taken place in the last few millennia (over and above the small crater-forming examples already mentioned). It is just that, for some reason, most people who study the past have chosen to avoid, or ignore, the issue. (Baillie 2006, 133–134)

Along with the science, Baillie cites contemporary evidence – some of this evidence has been relegated to 'myth' – from around the globe that indicate that the earth was, indeed, subjected to bombardment from space during the 14th century and that this may very well have been not only the cause of the 25 January 1348 earthquake, but also the cause of the Black Death. Baillie quotes a great selection of material from contemporary accounts including the work of Ziegler cited above: *"droughts, floods, earthquakes, locusts, subterranean thunder, unheard of tempests, lightning, sheets of fire, hail stones of marvelous* size, fire from heaven, stinking smoke, corrupted atmosphere, a vast rain of fire, masses of smoke" (Ziegler, paraphrased in Baillie 2006, 87). Ziegler discounts entirely reports of a black comet seen before the arrival of the epidemic but records: "heavy mists and clouds, falling stars, blasts of hot wind, a column of fire, a ball of fire, a violent earth tremor, in Italy a crescendo of calamity involving earthquakes, following which, the plague arrived" (Baillie 2006, 87).

As it happens, in the 1340s there was a veritable rash of earthquakes. In Rosemary Horrox's book, *The Black Death*, we find that a contemporary writer in Padua reported that not only was there a great earthquake on 25 January 1348, but it was at the twenty-third hour:

> In the thirty-first year of Emperoro Lewis, around the feast of the Conversion of St. Paul [25 January] there was an earthquake throughout Carinthia and Carniola which was so severe that everyone feared for their lives. There were repeated shocks, and on one night the earth shook 20 times. Sixteen cities were destroyed and their inhabitants killed …. Thirty-six mountain fortresses and their inhabitants were destroyed and it was calculated that more than 40,000 men were swallowed up or overwhelmed. (Horrox, quoted in Baillie 2006, 87)

The author goes on to say that he received information from "a letter of the house of Friesach to the provincial prior of Germany":

> It says in the same letter that in this year [… 1348 …] fire falling from heaven consumed the land of the Turks for 16 days; that for a few days it rained toads and snakes, by which many men were killed; that a pestilence has gathered strength in many parts of the world (Horrox, quoted in Baillie 2006, 88)

From Samuel Cohn's book:

> … a dragon at Jerusalem like that of Saint George that devoured all that crossed its path … a city of 40,000 … totally demolished by the fall from heaven of a great quantity of worms, big as a fist with eight legs, which killed all by their stench and poisonous vapours. (Cohn, quoted in Baillie 2006, 88)

A story by the Dominican friar Bartolomeo:

> … massive rains of worms and serpents in parts of China, which devoured large numbers of people. Also in those parts fire rained from Heaven in the form of snow (ash), which burnt mountains, the land, and men. And from this fire arose a pestilential smoke that killed all who smelt it within twelve hours, as well as those who only saw the poison of that pestilential smoke. (Cohn, quoted in Baillie 2006, 88)

Cohn writes:

> Nor were such stories merely the introductory grist of naïve merchants and pos-
> sibly crazed friars ...[even]... Petrarch's closes friend, Louis Sanctus, before em-
> barking on his careful reporting of the plague ... claimed that in September
> floods of frogs and serpents throughout India had presaged the coming to Europe
> in January of the three pestilential Genoese galleys ...[even]... the English chron-
> icler Henry Knighton ...[reported how]... at Naples the whole city was destroyed
> by earthquake and tempest.
>
> Numerous chroniclers reported earthquakes around the world, which prefig-
> ured the unprecedented plague. Most narrowed the event to Vespers, 25 January
> 1348. ...
>
> Of these earthquakes that "destroyed many cities, towns, churches, monasteries,
> towers, along with their people and beasts of burden, the worst hit was Villach
> in southern Austria. Chroniclers in Italy, Germany, Austria, Slavonia, and Poland
> said it was totally submerged by the quake with one in 10 surviving. (Cohn,
> quoted in Baillie 2006, 88–89)

A continental text dated Sunday 27 April 1348 states:

> They say that in the three months from 25 January [1348] to the present day, a total
> of 62,000 bodies were buried in Avignon. (Horrox, quoted in Baillie 2006, 98)

A German treatise unearthed by Horrox says:

> Insofar as the mortality arose from natural causes its immediate cause was a cor-
> rupt and poisonous earthy exhalation, which infected the air in various parts of
> the world .... I say it was the vapour and corrupted air which has been vented –
> or so to speak purged – in the earthquake that occurred on St. Paul's day [1348],
> along with the corrupted air vented in other earthquakes and eruptions, which
> has infected the air above the earth and killed people in various parts of the
> world. (Horrox, quoted in Baillie 2006, 99)

As Baillie notes, if this oft-cited earthquake was, in reality, the result of
cometary impacts then the corrupted air could be from one or two causes:
high-energy chemical transformations in the atmosphere or outgassings from
the earth itself.

The German historian, J. Hecker, informs us:

> On the island of Cyprus, the plague from the East had already broken out; when
> an earthquake shook the foundations of the island, and was accompanied by so

frightful a hurricane, that the inhabitants ... fled in dismay .... The sea overflowed
.... Before the earthquake, a pestiferous wind spread so poisonous an odour that
many, being overpowered by it, fell down suddenly and expired in dreadful ag-
onies ... and as at that time natural occurrences were transformed into miracles,
it was reported that a fiery meteor, which descended on the earth far in the East,
had destroyed everything within a circumference of more than a hundred leagues,
infecting the air far and wide. (Hecker, quoted in Baillie 2006, 100)

Jon Arrizabalaga compiled a selection of writings in an attempt to understand what educated people were saying about the Black Death while it was
happening. Regarding the terms used by doctors and other medical professionals in 1348 to describe the plague, he writes:

One ... Jacme d'Agramaont, discussed it in terms of an 'epidemic or pestilence
and mortalities of people' ... which threatened Lerida from 'some parts and regions neighbouring to us' .... Agramont said nothing concerning the term
epidímia, but he extensively developed what he meant by *pestilència*. He gave this
latter term a very peculiar etymology, in accordance with a form of knowledge
established by Isidore of Seville (570–636) in his *Etymologiae*, which came to be
widely accepted throughout Europe during the Middle Ages. He split the term
pestilència up into three syllables, each having a particular meaning: *pes* (= *tem-
pesta*: 'storm', 'tempest'), *te* (= *temps*: 'time'), and *lència* (= *clardat*: 'brightness',
'light'); hence, he concluded, the *pestilència* was 'the time of tempest caused by
light from the stars.' (Arrizabalaga, quoted in Baillie 2006, 102)

Historical evidence for recent earth-changing comet impacts is profuse.

As it happens, Isidore of Seville lived not long after another period of cometary bombardment over Europe that is also evident in the tree-ring and ice-core studies.

On 17 August 1999, the Knight Ridder Washington Bureau published an article by Robert S. Boyd entitled 'Comets may have caused Earth's great empires to fall', which included the following:

> Analysis of tree rings shows that in 540 AD in different parts of the world the climate changed. Temperatures dropped enough to hinder the growth of trees as widely dispersed as northern Europe, Siberia, western North America, and southern South America.
>
> A search of historical records and mythical stories pointed to a disastrous visitation from the sky during the same period, it is claimed. There was one reference to a "comet in Gaul so vast that the whole sky seemed on fire" in 540–41.
>
> According to legend, King Arthur died around this time, and Celtic myths associated with Arthur hinted at bright sky Gods and bolts of fire.
>
> In the 530s, an unusual meteor shower was recorded by both Mediterranean and Chinese observers. Meteors are caused by the fine dust from comets burning up in the atmosphere. Furthermore, a team of astronomers from Armagh Observatory in Northern Ireland published research in 1990 which said the Earth would have been at risk from cometary bombardment between the years 400 and 600 AD. ...
>
> Famine followed the crop failures, and hard on its heels bubonic plague that swept across Europe in the mid-6th century. ...
>
> At this time, the Roman emperor Justinian was attempting to regenerate the decaying Roman empire. But the plan failed in 540 and was followed by the Dark Ages and the rise of Islam.

There is a large body of material from that period which consistently points to a 'corrupted atmosphere, breathe the air and you die', and somehow, the ocean was involved as well as earthquakes and comets and fireballs in the sky. A report of the Medical Faculty of Paris prepared in October 1348 says:

> Another possible cause of corruption, which needs to be borne in mind, is the escape of the rottenness trapped in the center of the earth as a result of earthquakes – something that has indeed recently occurred. (quoted in Baillie 2006, 94)

In short, the French were aware of a series of earthquakes at the time that may have been caused by cometary impacts. One report of that period says that an earthquake lasted for six days and another claimed the period was ten days. Such events could also produce outgassing of all sorts of potentially deadly chemicals. Consider the following from Wikipedia's entry on the Lake Nyos Gas Explosion in Cameroon, 1986:

> Although a sudden outgassing of CO_2 had occurred at Lake Monoun in 1984, killing 37 local residents, a similar threat from Lake Nyos was not anticipated. However, on August 21, 1986, a limnic eruption occurred at Lake Nyos which triggered the sudden release of about 1.6 million tonnes of CO_2. The gas rushed down two nearby valleys, displacing all the air and suffocating up to 1,800 people within 20 km of the lake, mostly rural villagers, as well as 3,500 livestock. About 4,000 inhabitants fled the area, and many of these developed respiratory problems, burns, and paralysis as a result of the gases.
>
> It is not known what triggered the catastrophic outgassing. Most geologists suspect a landslide, but some believe that a small volcanic eruption may have occurred on the bed of the lake. ...
>
> It is believed that up to a cubic kilometre of gas was released. Because CO_2 is denser than air, the gas flowed off the mountainous flank in which Lake Nyos rests and down two adjoining valleys in a layer tens of metres deep, displacing the air and suffocating all the people and animals before it could dissipate. The normally blue waters of the lake turned a deep red after the outgassing, due to iron-rich water from the deep rising to the surface and being oxidised by the air. The level of the lake dropped by about a metre, representing the volume of gas released. The outgassing probably also caused an overflow of the waters of the lake. Trees near the lake were knocked down.

One wonders if similar events could be triggered by cometary impacts and if outgassing from oceans could be as dangerous and deadly? One also wonders, considering the fact that trees were 'knocked down', if this outgassing might not have been an impact event?

Baillie takes us through the science with hard numbers and graphs and shows us how these things were spoken about plainly by those who experienced the Black Death but, for some reason, modern historians all think these remarks about rains of fire and death and air that could kill were all just metaphors for a horrible disease. In the end, it is the science that must win on this one because totally independent researchers studying comets, tsunamis, carbon dioxide, ice cores and tree rings all observe in their data something very strange happening globally around the time that the Black Death decimated the human population of the earth.

Baillie notes in his closing remarks, worth repeating here:

> It is increasingly evident that intellectually the world is divided into two. There are those who study the past, in the fields of history and archaeology, and see no

evidence for any human populations ever having been affected by impacts from space. In diametric opposition to this stance there are those who study the objects that come close to, and sometimes collide with, this planet. Some serious members of this latter group have no doubt whatsoever that there must have been numerous devastating impacts in the last five millennia; the period of human civilization. (Baillie 2006, 199)

And yet, nobody talks about it.

Baillie presents sufficient data in his book to support the theory that the Black Death was due to localized, multiple impacts by comet debris – similar to the impacts on Jupiter by the fragments of comet Shoemaker-Levy back in 1994. As to exactly how these deaths occurred, there are a number of possibilities: earthquakes, floods (tsunami), rains of fire, chemicals released by the high-energy explosions in the atmosphere, including ammonium and hydrogen cyanide, and possibly even comet-borne disease pathogens.

If it has happened as often as Baillie suggests, it can happen again. And if, as we suspect, the earth is slated for a bombardment in the not-too-distant future, it seems that there are more ways to die in such an event than simply getting hit by a comet fragment.

3. Halloween and Ancient Techno-Spirituality

Before I get to the next interesting period of history alluded to by Clube – the Hundred Years War – I need to include some more background material. Specifically, I want to talk about Halloween. When you think of Halloween, what is the first image that comes to mind?

When I think of Halloween, I think of grade-school art projects where we cut out silhouettes of witches to paste onto large yellow moons made of construction paper. The witch was always on a broom with her black dress flying in the wind, accompanied by a black cat sitting on the back of the broom.

In a significant way, Halloween is associated with witches, evil women who consort with the devil and do evil things like caging lost children to fatten them up and eat them, giving poisoned apples and setting up spinning wheels to poison abandoned or hapless princesses who are only looking for true love.

The word 'witch' comes to us from the Old English *wicca*, which was a masculine word meaning 'wizard'. The feminine version was *wicce*, pronounced 'witch'. This came from the Middle High German *wicken*, which meant 'to bewitch', and even older, from Old High German *wih* which meant 'holy'. The dictionary tells us that a witch is someone who has malignant supernatural powers and practices spell-casting with the aid of a devil or familiar. It also refers to an ugly old woman, or a beautiful young woman. The word 'witch' is an epithet for any woman who isn't inclined to be a doormat, flung to the floor by any individual who wants her to be subject to his or her will. Last of all, a witch is a practitioner of Wicca.

Wicca is a British construct created by an amateur anthropologist named Gerald Gardner who claimed to have had many interesting encounters and experiences with the occult and paranormal throughout his life. At one point, he claimed to have doctoral degrees from the Universities of Singapore and Toulouse, which was a lie. He claimed that he was initiated into a New Forest coven of witches that was the survival of a pre-Christian pagan witch cult. This alleged ancient coven has been shown by subsequent research to have been formed in the early 20th century and its ideas were based mainly on folk magic and the theories of Margaret Murray, so again, Gardner's honesty is rather suspect.

Gardner incorporated elements from Freemasonry, ceremonial magic, and the imaginings of Aleister Crowley and others. Most of what one sees when carefully examining these elements that combined to form modern Wicca bears no relationship whatsoever to the ancient religions as they can be discerned by deep study. Rather, these elements are likely more influenced by taking the descriptions of the persecutors of witches during the Inquisition as a guideline, instead of realizing that they were the defamatory falsifications of psychopaths.

It is more likely that those accused of witchcraft during the witch persecutions were following beliefs akin to those of the Cathars – dualism – or even more ancient dualistic concepts. They also likely employed ancient knowledge handed down from Paleolithic shamanic systems that had little to nothing to do with 'ceremonial magic', spells or a 'liberal code of morality'. Unfortunately, neither Gardner nor Crowley had access to modern scientific archaeological studies from which one can actually infer something about the abilities, beliefs and practices of our truly remarkable ancestors.

My main line of work is all about following the lines of pagan/shamanistic ideas and teachings back to the Ice Ages – the cave painters, the Northern European origins – to find the most original, fundamental, common foundation of all of them. The idea that there was a time when man was directly in contact with the Celestial Beings is at the root of many of the myths of the Golden Age. Myths tell us of a time when 'the gods withdrew' from mankind. As a result of some 'happening', i.e., 'The Fall', the communications were broken off and the Celestial Beings withdrew to the highest heavens.

But the myths also tell us that there were still certain people who were able to 'ascend' and commune with the gods on behalf of their tribe or family. Through them, contact was maintained with the 'guiding spirits' of the group. The beliefs and practices of present-day shamans are a survival of a profoundly modified and even corrupted and degenerated remnant of this archaic technology of concrete communications between heaven and earth. This shamanism seems to have been born in Western Europe with the arrival of Cro-Magnon man and the myths seem to have been redacted repeatedly until we have numerous claims of occult secrets of various sorts revived by this or that person or group, including Wicca. If that is the case, then true 'witchcraft' is really shamanism, aka Druidism, and even more, as we shall see. Mircea Eliade writes:

> Recent researches have clearly brought out the "shamanic" elements in the religion of the paleolithic hunters. Horst Kirchner has interpreted the celebrated re-

lief at Lascaux as a representation of a shamanic trance. … Finally, Karl J. Narr has reconsidered the problem of the "origin" and chronology of shamanism in his important study … He brings out the influence of notions of fertility ("Venus statuettes") on the religious beliefs of the prehistoric North Asian hunters; but this influence did not disrupt the Paleolithic tradition. … it is in this "Vorstell-ungswelt" that the roots of the bear ceremonialism of Asia and North America lie. Soon afterward, probably about 25,000 [BCE], Europe offers evidence for the earliest forms of shamanism (Lascaux) with the plastic representations of the bird, the tutelary spirit, and ecstasy.

… What appears to be certain is the antiquity of "shamanic" rituals and symbols. It remains to be determined whether these documents brought to light by prehistoric discoveries represent the first expressions of a shamanism in *statu nascendi* or are merely the earliest documents today available for an earlier religious complex, which, however, did not find "plastic" manifestations (drawings, ritual objects, etc.) before the period of Lascaux.

… It is indubitable that the celestial ascent of the shaman is a survival, profoundly modified and sometimes degenerate, of this archaic religious ideology centered on faith in a celestial Supreme Being and belief in concrete communications between heaven and earth.

… The myths refer to more intimate relations between the Supreme Beings and shamans; in particular, they tell of a First Shaman, sent to earth by the Supreme Being or his surrogate to defend human beings against diseases and evil spirits. (Eliade 1951, 503–504)

It was in the context of the withdrawal of the Celestial Being that the meaning of the shaman's ecstatic experience changed. Formerly, the activity was focused on communing with the god and obtaining benefits for the tribe. The shift of the function of the shaman associated with the withdrawal of the benevolent god/goddess was to 'battling with evil spirits and disease'. This is a sharp reminder of the work of Jesus, healing the sick and casting out demons – the shamanic exemplar 'after the Fall'.

There was, it seems, another consequence of this shift. Increasingly, the descents into the 'underworld' and the relations with 'spirits' led to their 'embodiment' or in the shaman's being 'possessed by spirits'. What is clear is that these were innovations, most of them recent. What is particularly striking in the research of the historiographers of myth, legend and shamanism is the discovery of the "*influences from the south, which appeared quite early and which altered both cosmology and the mythology and techniques of ecstasy*". Among these southern influences were the contribution of Buddhism and Lamaism, added to the Iranian and, in the last analysis, Mesopotamian influences that preceded them.

> ... the initiatory schema of the shaman's ritual death and resurrection is likewise an innovation, but one that goes back to much earlier times; in any case, it cannot be ascribed to influences from the ancient Near East. But the innovations introduced by the ancestor cult particularly affected the structure of this initiatory schema. The very concept of mystical death was altered by the many and various religious changes effected by lunar mythologies, the cult of the dead, and the elaboration of magical ideologies. (Eliade 1964, 506)

It is clear that shamanism, as it is known, has declined from its original unified and coherent system. One reason for thinking so is that, while there are many local terms for a male shaman, there is only one for a female shaman. Shamanism, it seems, was formerly a woman's activity. In one Tartar dialect, *utygan*, the word for a woman-shaman, also means 'bear'.

> ... the magico-religious value of intoxication for achieving ecstasy is of Iranian origin. ... concerning the original shamanic experience ... [n]arcotics are only a vulgar substitute for "pure" trance. The use of intoxicants ... is a recent innovation and points to a decadence in shamanic technique. Narcotic intoxication is called on to provide an *imitation* of a state that the shaman is no longer capable of attaining otherwise. Decadence or ... vulgarization of a mystical technique – in ancient and modern India, and indeed all through the East, we constantly find this strange mixture of "difficult ways" and "easy ways" of realizing mystical ecstasy or some other decisive experience. (Eliade 1964, 401)

Now, let me make a point here. The religion of the Ice Age was so satisfying to all the peoples of the earth that it was stable for over 25,000 years, as is evidenced by the archaeological and historical data. There were shamans – women – who engaged in ecstatic ascents that brought benefits to the tribe and, later, defended the tribe against negative influences. In short, it seems that paganism, even Druidism, was in a sense the original Christianity, and the original 'Christed Ones' were women.

Many researchers repeatedly point out that Christianity has pagan roots. I would say that this is true; more than anybody suspects. And if the lines of research I have presented in my book, *The Secret History of the World*, are anything to go by, then the original 'witches' were Christs.

This, of course, leads us to wonder how can things get so turned around that we actually end up believing the opposite of the truth in almost every field of endeavor? We may turn away from mainstream religions that we can see are false and contradictory, only to fall into the arms of New Age religions that are not any better, being just another variation on a control system designed to prevent us from accessing what is real.

A female shaman from the Altai mountains.

Now, back to Halloween. The last day of October is a holiday that is said to be the ancient Celtic celebration of the 'end of summer', Samhain, Halloween, or All Hallows Eve. As I mentioned above, many people think of witches when you say the word 'Halloween'. But why is October 31st associated with witches and celebrated as the 'end of summer' when the Autumnal equinox, over a month earlier, is the actual end of summer?

According to British historian Ronald Hutton, the festival of Samhain celebrates the end of the 'lighter half' of the year and beginning of the 'darker half' and is sometimes regarded as the Celtic New Year. According to folklorist John Gregorson Campbell and archaeologist Bettina Arnold, the ancient Celts believed that the curtain separating this world from the Otherworld became thin on Samhain, allowing spirits (both good and bad) to easily traverse the otherwise sturdy barrier. They dealt with this by inviting the good spirits in – usually family ancestors – and utilizing various techniques to ward off or scare away any bad spirits. It is suggested that this is the origin of wearing costumes disguising oneself as skeletons, ghosts, and goblins, the principle being that if you looked horrible enough, you could even scare away the devil himself!

Samhain was also the time when people in the old times took stock of their food supplies, butchered cattle and pigs, and prepared grains and other foodstuffs for winter storage.

Bonfires were an important part of the celebrations. Hearth fires were put out, the bones of the slaughtered cattle were tossed into the bonfire, and each home re-lit their hearth fire from the coals of the bonfire. Sometimes two bonfires would be built so that people could pass between them with their livestock for 'purification'. This practice may be a survival of the times when the ancient tribes purified themselves by burning alive: *a)* any members who were less than perfect so that the tribe could be cleansed of sinful elements, or *b)* those members who were actually perfect in some way and volunteered to be offered as a sacrifice to appease the gods so the rest of the tribe could live in peace for another year. This is, in fact, an interesting clue.

The name 'Halloween' is an old Scottish variant of 'All Hallows Eve', or the night before All Hallows Day, or the Feast of All Saints. What is interesting to observe here are the old customs regarding this day, and especially the following two days, from around the world that were later Christianized, but obviously represent something far more ancient.

In Portugal and Spain, offerings are made on All Saints Day. In Mexico, All Saints coincides with the celebration of the Day of the Innocents, part of the Day of the Dead honoring deceased children and infants. In Portugal, children go door-to-door where they receive cakes, nuts and pomegranates. The holiday focuses on family gatherings where prayers for, and remembrance of, friends and family that are departed are the focus. Traditions include building altars honoring the deceased, feasting on sugar skulls (devouring death?) and the favorite foods and beverages of the departed, decorating with marigolds and visiting graves with these as gifts. Scholars trace the origins of the modern holiday to indigenous observances dating back thousands of years and to an Aztec festival dedicated to a goddess called Mictecacihuatl, the Queen of Mictlan, or the underworld. It was believed that she was sacrificed as an infant, and she is represented with a de-fleshed body, and her gaping jaw swallows the stars during the day.

In the Philippines, this day is called *Undas, Todos los Santos* (literally 'All Saints'), and sometimes *Araw ng mga Namayapa* (approximately 'Day of the Deceased'). This day and the one before and one after it are spent visiting the graves of deceased relatives, where prayers and flowers are offered, candles are lit and the graves themselves are cleaned, repaired and repainted. The practices are similar in most European countries.

In Brazil, *Dia de Finados* is a public holiday that many Brazilians celebrate by visiting cemeteries and churches. In Spain, there are festivals and parades, and, at the end of the day, people gather at cemeteries and pray for their dead loved ones. Similarly themed celebrations appear in many Asian and African cultures.

These celebrations, which occur on November 1st and 2nd, and have indigenous forms that the Church assimilated, strike us as curious. It seems that what is important is that *they follow immediately on the heels of October 31st*. One is compelled to ask why. What happened on October 31st that turned the following day into the Day of the Dead?

The symbols associated with Halloween formed over time and, just as the medieval Church assimilated the ancient death-themed images and practices, many of the customs of contemporary times have assimilated the medieval practices. In traditional Celtic Halloween festivals, large turnips were hollowed out, carved with faces and placed in windows to ward off evil spirits. The American tradition of carving pumpkins was originally associated with harvest time in general, not becoming specifically associated with Halloween until the mid-to-late 1800s.

While most Christians just think of Halloween as a secular holiday which allows children to dress up in silly costumes, eat candy, and generally make fun of everything that is normally scary in our world, some other – mostly fundamentalist – Christians ascribe a negative influence to the celebration because they feel it celebrates paganism and the occult, or trivializes it so that their members are not properly fearful of ghosts, demons and the devil. Jehovah's Witnesses do not celebrate Halloween because they believe true Christians should not celebrate anything that originated from a pagan holiday. This is ironic considering what I have written above about original Christianity. How did we get from there – true spirituality that honored women, with women shamans that provided for the tribe – to here, the modern-day Christian view of women as something barely human?

Many of those who follow pagan ways consider the season to be a holy time of year and, naturally, Wiccans feel that the whole holiday as it is generally celebrated, is offensive because it associates witches with the other list of 'evil spirits' that need to be warded off. They are right about that, but most of what they consider 'Wicca' is as wrong as Christianity is wrong.

This brings us back to the question this chapter hopes to answer: What is the origin of Halloween, what does it really commemorate, and why are witches associated with it?

The first point I would like to bring up is that I think, when we look at Halloween, we are seeing something very ancient that is filtered through many layers of interpretation. What is consistent throughout, however, is the theme of easy traversal of the border between life and death, leading mainly to death, which suggests that death on a massive scale came on Halloween a very long time ago. Whatever it was, it was so terrifying, so widespread, that cultures the world over have commemorated it, and the days following it, in ways that appear to be designed to ward it off, to prevent it from ever happening again. And along the way, things happened that turned everything around so that those individuals – real, holy witches – who actually might be capable of knowing such things, of ameliorating such terrors, became identified with the cause of the death and destruction.

In his book, *The Worship of the Dead, or the Origin and Nature of Pagan Idolatry and Its Bearing Upon the Early History of Egypt and Babylonia*, John Garnier writes that the modern-day celebrations for the dead focused around All Hallows Eve, including the following few days, originated to memorialize the people who died in the Deluge brought by God on a wicked world. He bases this on Genesis 7:11.

He writes:

There is hardly a nation or tribe in the world which does not possess a tradition of the destruction of the human race by a flood; and the details of these traditions are too exactly in accordance with each other to permit the suggestion, which some have made, that they refer to different local floods in each case.

The mythologies of all the ancient nations are interwoven with the events of the Deluge and are explained by it, thereby proving that they are all based on a common principle, and must have been derived from a common source.

It is clear from these remarks that one or other of the two great events in the history of the Deluge, namely, the commencement of the waters and the beginning of their subsidence, were observed throughout the ancient world, some nations observing one event and some the other.

It would also appear probable that the observance of this festival was intimately connected with, and perhaps initiated, that worship of the dead which, as we shall see, was the central principle of the ancient idolatry.

The force of this argument is illustrated by the fact of the observance of a great festival of the dead in commemoration of the event, not only by nations more or less in communication with each other, but by others widely separated, both by the ocean and by centuries of time. This festival is, moreover, held by all on or about the very day on which, according to the Mosaic account, the Deluge took place, viz. the seventeenth day of the second month – the month nearly corresponding with our November. (Garnier 1904, 3–11)

I don't know which of the many Jewish calendars he was using, but Garnier's point was that holidays that bring honor to dead spirits are un-Christian because they have pagan roots (never mind all the honoring of dead saints and praying to them – they were Christian *before* they died, or so it is claimed) and because they are founded on honoring the deaths of the wicked people who were justifiably destroyed by God in Noah's Flood. This 'Christian' spin on all things pagan is why, apparently, Halloween has such an emphasis on demonic images, ghosts, monsters, and gruesome things in general, because, as Garnier points out, the Flood meant the death of the hybrid children of demons, the Nephilim (see Gen. 6:1–4, 13 and the Book of Enoch).

So, based on the above, the idea that Halloween represents a memory of the Flood seems to be just a conjecture made by a religious antiquarian of olden times; nothing to see here. But, maybe not? Maybe Garnier was onto something and didn't really know what it was?

The Flood of Noah and Cometary Bombardment

Regarding the alleged Flood of Noah, we can say that at more than one point in our known history, civilizations and cultures have collapsed and/or disappeared or been destroyed by no-one-knows-what. The Akkadian Empire in Mesopotamia, the Old Kingdom in Egypt, the Early Bronze Age civilization in Palestine, Anatolia and Greece, as well as the Indus Valley civilization in India, the Hilmand civilization in Afghanistan and the Hongshan in China, all fell into ruin at more or less the same time. Not long afterward, in archaeological time (though the chronology is a mess), destruction came to the Myceneans of Greece, the Hittites of Anatolia, the Egyptian New Kingdom, Late Bronze Age Palestine, and the Shang Dynasty of China.

Researchers in the fields of archaeology and history are baffled by the lack of any direct archaeological or written explanations for the causes (as opposed to the effects), though there is a rich body of myth and folklore that very well might provide the answers if analyzed correctly. Since the 'experts' in those fields have consigned myth to superstition, while simultaneously believing that the historicized myths incorporated into the Bible are history, they aren't getting very far with their problem and usually ascribe the collapse of civilizations to invasion and warfare on a gargantuan scale.

Some decades ago, certain natural scientists became intrigued by the problem and, concentrating on the Bronze Age collapses listed above, realized that the range of evidence suggested natural causes rather than human actions like invasion and warfare. So, they all started talking about climate change, volcanic activity, and earthquakes. At present, these types of explanations are actually included in some of the standard historical accounts of the Bronze Age period, though many problems still remain: no single explanation appeared to account for all the evidence.

Immanuel Velikovsky upset everyone by suggesting that the Exodus – but *only* the Exodus – was caused by a bombardment of rocks, dust, carbons, and so on, as a result of Venus running amok in the solar system. He collected an amazing assortment of myths and legends from around the world that strongly suggested that some sort of global cataclysm was being described, but when, where and how exactly it happened was rather sketchy. There were others who wrote and talked about these matters before Velikovsky, including Ignatious Donnelly, who deserves an honorable mention for ascribing the myths to the Great Flood of Noah, which he claimed was actually the destruction of Atlantis as described by Plato. Whether or not there was an advanced civilization known as Atlantis is not our concern here, but whether or not there was a flood, and when it may have occurred, is.

In the late 1970s, British astronomers Victor Clube and Bill Napier of Oxford

University began investigating cometary impact as the ultimate cause. In 1980, Nobel Prize winning physicist Luis Alvarez and his colleagues published a paper in *Science*, which argued that a cosmic impact is what led to the extinction of the dinosaurs. Alvarez's paper had immense influence, though that influence acted in different ways on the two sides of the Atlantic, as we learned above. In the US, there is the 'wishful thinking' school that posits that only asteroid impacts are significant and they are so rare that we don't have to worry. In Britain, further research by astronomers Clube and Napier, Prof. Mark Bailey of the Armagh Observatory, Duncan Steel of Spaceguard Australia, and Britain's best-known astronomer Sir Fred Hoyle, all led to their support of the theory of cometary impact, loosely termed the 'British School of Coherent Catastrophism'.

According to Clube and Napier, *et al.*, in the same way that Jupiter was struck repeatedly in 1994 by the million-megaton impacts of the comet Shoemaker-Levy, so earth was bombarded 13,000 years ago by the fragments of a giant comet that broke up in the sky before the terrified eyes of humanity. The multiple impacts on the rotating planet caused tidal waves, raging fires, atomic bomb-like blasts, the mass extinction of many prehistoric species such as the mammoth and sabre-toothed tiger, most of humanity, and left the world in darkness for months. (See: *The Cosmic Serpent* and *The Cosmic Winter* by Clube and Napier, as well as *The Origin of the Universe and the Origin of Religion – Anshen Transdisciplinary Lectureships in Art, Science, and the Philosophy of Culture* by Fred Hoyle.)

Some American scientists are joining the Coherent Catastrophism group, as well. Physicist Richard Firestone and geologists Allen West and Simon Warwick-Smith write in their book, *The Cycle of Cosmic Catastrophes*:

In 1990, Victor Clube, an astrophysicist, and Bill Napier, an astronomer, published *The Cosmic Winter*, a book in which they describe performing orbital analyses of several of the meteor showers that hit Earth every year. Using sophisticated computer software, they carefully looked backward for thousands of years, tracing the orbits of comets, asteroids, and meteor showers until they uncovered something astounding. Many meteor showers are related to one another, such as the Taurids, Perseids, Piscids, and Orionids. In addition, some very large cosmic objects are related: the comets Encke and Rudnicki, the asteroids Oljato, Hephaistos, and about 100 others. *Every one of those 100-plus cosmic bodies is at least a half-mile in diameter and some are miles wide.* And what do they have in common? According to those scientists, *every one is the offspring of the same massive comet that first entered our system less than 20,000 years ago!* Clube and Napier calculated that, to account for all the debris they found strewn throughout our solar system, *the original comet had to have been enormous.*

... Clube and Napier also calculated that, because of subtle changes in the orbits of Earth and the remaining cosmic debris, Earth crosses through the densest part of the giant comet clouds about every 2,000 to 4,000 years. When we look at climate and ice-core records, we can see that pattern. For example, the iridium, helium-3, nitrate, ammonium, and other key measurements seem to rise and fall in tandem, producing noticeable peaks around 18,000, 16,000, 13,000, 9,000, 5,000, and 2,000 years ago. In that pattern of peaks every 2,000 to 4,000 years, *we may be seeing the "calling cards" of the returning mega-comet.*

Fortunately, the oldest peaks were the heaviest bombardments, and things have been getting quieter since then, as the remains of the comet break up into even smaller pieces. The danger is not past, however. Some of the remaining miles-wide pieces are big enough to do serious damage to our cities, climate, and global economy. Clube and Napier (1984) predicted that, *in the year 2000 and continuing for 400 years, Earth would enter another dangerous time in which the planet's changing orbit would bring us into a potential collision course with the densest parts of the clouds containing some very large debris.* Twenty years after their prediction, we have just now moved into the danger zone. It is a widely accepted fact that some of those large objects are in Earth-crossing orbits at this very moment, and the only uncertainty is whether they will miss us, as is most likely, or whether they will crash into some part of our planet. [emphasis added] (Firestone *et al.* 2006, 354–355)

And so we see that this new type of 'natural disaster' is beginning to be regarded by many scholars as the most probable single explanation for widespread and simultaneous cultural collapses at various times in our history. These ideas have been advanced largely by astronomers, geologists, dendrochronologists, and other scientists, and remain almost completely unknown among archaeologists and historians, which significantly hampers their efforts to explain what they may be seeing in the historical record.

The new theory posits trains of cometary debris that repeatedly encounter the earth. We know most of these trains as meteor showers – tiny particles of cosmic material whose impact is insignificant. Occasionally, however, in these trains of debris, there are chunks measuring between one and several hundred meters in diameter. When these either strike the earth or explode in the atmosphere, there can be catastrophic effects on our ecological system. Multimegaton explosions of fireballs can destroy natural and cultural features on the surface of the earth by means of tidal-wave floods (if the debris lands in the sea), fire blasts and seismic damage leaving no crater as a trace, just scorched and blasted earth. In the case of a significant bombardment, an entire small country could be wiped out, completely vaporized.

The Tunguska event (mentioned briefly above) occurred in 1908 over

Siberia, when a bolide exploded about 5 km above ground and completely devastated an area of some 2,000 km² through fireball blasts. This cosmic body, thought to have measured only 60 m across, had the impact energy of about 20 to 40 megatons, and was equivalent to the explosion of about 2,000 Hiroshima-size nuclear bombs, even though there was no actual physical impact on the earth. In other words, if there were ancient, advanced civilizations, and if they were destroyed by multiple Tunguska-like events, it is no wonder there is no trace, or at least very little, which is usually ascribed to 'anomaly' in the final analysis.

For years, the astronomical mainstream was highly critical of Clube and Napier and their giant comet hypothesis. However, the impacts of comet Shoemaker-Levy 9 on Jupiter in 1994 led to a rather rapid turnaround in attitude. The comet, watched by the world's observatories, was seen to split into 20 pieces and slam into different parts of the planet over a period of several days. A similar event vis-à-vis our planet would have been devastating, to understate the matter. In recent times, the increasing numbers of fireballs and comets, the fact that Jupiter has been impacted yet again and again in 2010, suggests to us that Victor Clube and Bill Napier are correct: we are living in a very dangerous period.

Trees were knocked down and burned over hundreds of square kilometers by the Tunguska meteoroid impact.

As mentioned in the previous chapter, the work of John Lewis brings to the table the idea that some well-known historical earthquakes could very well have been impact events. The dates that these researchers have given to events that can be discerned in the scientific records are 12,800, 8,200, 5,200, and 4,200 BP ('years before the present'). These can be adjusted as more precise dating methods are developed or applied.

The 12,800 BP event is the one of most interest because it is the one which, apparently, nearly destroyed all life on earth. At the very least, it destroyed the mega-fauna on all continents. Plato wrote about the catastrophic destruction of Atlantis that occurred in a day and a night about 11,600 years ago, which is pretty darn close. This event is the topic Firestone, West and Warwick-Smith cover exhaustively in their book *The Cycle of Cosmic Catastrophes*. They include a great many Native American myths that describe the event side by side with their own scientific work on the evidence.

As already mentioned, Clube and Napier identified the progenitor of the Taurid complex as a giant comet that was thrown into a short-period (about 3.3-year) orbit sometime in the last twenty to thirty thousand years. The Taurid complex currently includes the Taurid meteor stream, comet Encke, 'asteroids' such as 2101 Adonis and 2201 Oljato, and enormous amounts of space dust. Asteroids in the Taurid complex appear to have associated meteor showers, which means that many asteroids are likely to be extinct comets. In other words, there can be more than just dust and snow in a comet – there can be a significant rocky core and lots of poisonous gasses and chemicals as well.

We come now to the bit of evidence that may link these comet phenomena and Halloween. As it happens, the end of June and the end of October/beginning of November are the times when the earth passes through the Taurid stream. That means that the event that marked the boundary between the Pleistocene and Holocene (present epoch) must have occurred at the end of October. It was a day when the boundaries between the living and the dead became very thin, because nearly every living thing on this planet perished and the memory of this event has come down to us in the 'end of summer' commemoration we call Halloween, known in the Bible as the Flood of Noah.

Clube and Napier write:

> … Meteor streams are fossil evidence of past intersections with comet orbits … the major streams are of great antiquity …
>
> The progenitor of comet Encke and the Taurids, supposing it to have been about 20 km in diameter, would, at its closest approaches to the Earth, have attained a magnitude -12, approaching that of the Moon and sufficient to throw shadows at night. It would have appeared as an intense yellow spot of light surrounded by a circular coma probably larger than the full Moon, with a tail stretch-

ing across a large part of the sky ... graduating from bluish white near the nucleus to a deep red in colour ... If the disintegration history revealed by the current debris took place within the sight of men, then there would have been occasions when subsidiary comets, perhaps even an array, would have been observed. ... There would (be) greatly enhanced seasonal fireball activity, rising to enormous levels at periodic intervals corresponding to a strong commensurability between the orbital periods of Earth and Encke; and the risk of Tunguska-like impacts would have been greatest. In a periodic orbit, the close approaches would obviously have been predictable. Indeed, if, at these close approaches, the Earth ran into debris of the sort we have discussed, prediction would have been a matter of urgency ...

The author of Genesis (15:17) wrote: 'When the sun went down, and it was dark, behold a smoking furnace and a burning lamp. ...' The description appears to be that of a comet; but its representation is that of a vision of God to Abraham. Or again, in I Chronicles (21:16): 'And David lifted up his eyes, and saw the angel of the Lord standing between the earth and the heaven, having a drawn sword in his hand stretched out over Jerusalem. Then David and the elders of Israel, who were clothed in sackcloth, fell upon their faces.' Once more the object is seen as a divine being, and 'angel of the Lord', and a religious interpretation is placed on a natural phenomenon. (Clube & Napier 1982, 154, 156)

Clube, Napier, Hoyle and others make a good case for the origins of Judaism in celestial phenomena, later twisted and distorted by priests into the superstition it is today. Christopher Knight and Robert Lomas wrote a fascinating book, *Uriel's Machine*, about the megalithic cultures wherein they propose that stone circles were constructed as astronomical observatories that were not for the purpose of knowing when to plant the corn, but rather to keep a watchful eye on errant comets. They make a very good case.

The beginnings of Christianity may have been the result of similar cosmic encounters. In his book, *A Myth of Innocence*, Burton Mack writes:

The story Josephus tells of the sixties is one of famine, social unrest, institutional deterioration, bitter internal conflicts, class warfare, banditry, insurrections, intrigues, betrayals, bloodshed, and the scattering of Judeans throughout Palestine. ... There were wars, rumors of wars for the better part of ten years, and Josephus reports portents, including a brilliant daylight in the middle of the night! (Mack 1988)

Romano-Jewish historian Josephus gives several portents of the evil to befall Jerusalem and the Temple. He described a star resembling a sword, a comet that *"continued a whole year..."*, a light shining in the Temple, a cow giving birth to a lamb at the moment it was to be sacrificed in the Jerusalem Temple, armies fighting in the sky, and a voice from the Holy of Holies declaring, *"We are departing"* (Josephus, *Jewish Wars*, 6). (Obviously, the voice was apocryphal.)

Some of these portents are mentioned by other contemporary historians, Tacitus for example. However, in book five of his *Histories*, Tacitus castigated the superstitious Jews for not recognizing and offering expiations for the portents to avert the disasters. He put the destruction of Jerusalem down to the stupidity or willful ignorance of the Jews themselves in not offering the appropriate sacrifices.

In short, it very well may be that the eschatological writings in the New Testament, the very formation of the myth of Jesus, were based on cometary events of the time, including a memory of the 'Star in the East'. The destruction of the Temple at Jerusalem may very well have been an 'act of God', as reported by Mark in his Gospel, though not quite as true believers think it was.

This brings us, of course, to the transition: the imposition of Christianity on Europe by Constantine. Paul K. Davis writes: *"Constantine's victory gave him total control of the Western Roman Empire, paving the way for Christianity as the dominant religion for the Roman Empire and ultimately for Europe."* (Davis 1999, 78)

It is commonly stated that on the evening of 27 October, with the armies preparing for battle, Constantine had a vision, which led him to fight under the protection of the Christian god. The details of that vision, however, differ depending on the source reporting it.

Lactantius, an early Christian writer of the time in question, states that, in the night before the battle, Constantine was commanded in a dream to *"delineate the heavenly sign on the shields of his soldiers"* (*On the Deaths of the Persecutors*, 44.5). He followed the commands of his dream and marked the shields with a sign denoting Christ. Lactantius describes that sign as a 'staurogram', or a Latin cross with its upper end rounded like a P. There is no certain evidence that Constantine ever used that sign, opposed to the better known Chi-Rho sign described by Eusebius, but it is certainly suggestive since it would look a bit like a mushroom cloud.

New Scientist (178:2400, 21 June 2003, 13) reported the discovery of a meteorite impact crater dating from the fourth or fifth century CE in the Apennines. The crater is now a 'seasonal lake', roughly circular, with a diameter of between 115 and 140 meters, which has a pronounced raised rim and no inlet or outlet and is fed solely by rainfall. There are a dozen much smaller craters nearby, such as would be created when a meteorite with a diameter of some 10 meters shattered during entry into the atmosphere.

A team led by the Swedish geologist Jens Ormo believes the crater was caused by a meteorite landing with a one-kiloton impact – equivalent to a very small nuclear blast – and producing shock waves, earthquakes *and a mushroom cloud.* Samples from the crater's rim have been dated to the year 312, but small amounts of contamination with recent material could account for a date significantly later than 312.

The legend of a falling star has been around in the Apennines since Roman times, but the event that it describes has been a mystery. Other accounts from the 4th century describe how barbarians stood at the gates of the Roman Empire while a Christian movement threatened its stability from within. The emperor Constantine saw an amazing vision in the sky, converted to Christianity on the spot, and led his army to victory under the sign of the cross. But what did he see?

Could the impact of a meteorite hitting the Italian Apennines, or a Tunguska-like overhead cometary explosion, have been the sign in the sky that encouraged Constantine to invoke the Christian God in his decisive battle in 312, when he defeated his fellow Emperor Maxentius at the Milvian Bridge?

The conversion of the emperor to Christianity certainly couldn't change the beliefs and practices of most of his subjects. But he could – and did – choose to grant favors and privileges to those whose faith he had accepted. He built churches for them, exempted the priesthood from civic duties and taxes, gave the bishops secular power over judicial affairs, and made them judges against whom there was no appeal. All of which sounds a lot like how a fascist regime takes over.

So, let's recap here: The god of the Jews leaped upon the stage of history – probably as a cometary event that was memorialized as the plagues in Egypt and recast as a heroic 'Exodus story'. Through his priests, centuries after the event when the reality of the 'god' was forgotten, this god promised his people something new and different – destruction of everybody else on the planet who were nasty to them – and only those who followed his rules carefully would survive and get to rule everybody else. Notice that this did not necessarily mean resurrection – it was to be a physical earthly kingdom with the Jews in top position.

Early Christianity had very distinct and novel ideas that were grafted onto Judaism. Christianity, in turn, retained and passed on in a virulent way certain ideals of Judaism, which have produced the foundation upon which our present culture is predicated.

'Pagan' Christianity

The main template of Christianity – received directly from Judaism – is that of sin. The history of sin from that point to now is a story of its triumph. Awareness of the nature of sin led to a growth industry in agencies and techniques for dealing with it. These agencies became centers of economic and military power, as they are today.

Christianity, promoting the ideals of Judaism under a thin veneer of the 'New Covenant', changed the ways in which men and women interacted with one another. It changed the attitude to life's one certainty: death. It changed the degree of freedom with which people could acceptably choose what to think and believe.

Pagan cults also dealt with the issues of suffering and troubles. The big difference was that, to the pagans, troubles fell on a person because they may have failed to propitiate the appropriate god or goddess. Suffering and troubles were a consequence of the actions of the gods – who were surprisingly human-like and fickle – and were not a personal, internal 'flaw' that damned the individual.

Another big difference between pagan cults and monotheistic cults was that pagans were not committed to revealed beliefs in the strong Christian sense. In other words, faith was neither endorsed nor encouraged. Pagans performed rites, but professed no creed or doctrine. The rites included detailed rituals involving the offering of animal victims to their gods, but there was nothing like the 'faith' of Judaism or Christianity.

The main template of Christianity is sin.

To be a follower of pagan religion, one did not have to accept the philosophic theology, nor did he have to belong to a 'mystery cult' where myth and ritual were closely entwined. These were just 'options'. What the myths actually did was confirm man's constant awareness of the potential anger of the gods, the uncertainties of Nature. Pausanias, a Greek geographer of the 2nd century AD, did not accept the outlandish stories of mythology. But there was one thing that Pausanias was sure about: *the tales of the past anger of a god, which had manifested in famines and earthquakes and cataclysm.* He reminds us of how fragile civilization is against the constant dangers of geology and the weather.

And so it was, to 'follow pagan religion' was essentially to accept this tradition of the past anger of the gods expressed in the violence of nature, and *that the gods could be appeased*. And it was precisely this fear of nature itself – of the gods that expressed themselves in the forces of nature – that caused the pagans to reject the Jews and Christians for claiming that they were immune to such things because their god had power over nature and would save them from calamity.

This brings us to another difference between the ancient myths and cults and Judaism, Christianity and Islam: where the pagan cults offered myths of their gods, Jews and Christians produced a recent, living *history*. The pagan cults had 'mysteries' to which very few – if anyone at all – had access. Monotheism offered a 'revelation' direct from God. Never mind that the history consisted of the plagiarized myths of other cultures that had been dressed in historical clothing as the 'History of Israel'.

Pagans had been intolerant of the Jews and Christians whose religions tolerated no gods but their own. The rising domination of Christianity created a much sharper conflict between religions, and religious intolerance – incepted by Christianity – became the norm, not the exception. Christianity brought the open coercion of religious belief. You could even say that, by the modern definition of a cult as a group that uses manipulation and mind control to induce worship, Christianity is the mother of all cults – in service to the misogynistic, fascist ideals of Judaism.

The rising Christian hierarchy of the Dark Ages was quick to mobilize military forces against believers in other gods and, most especially, against other Christians who promoted less fascist systems of belief. This probably included the original Christians and the original teachings. One wonders, of course, about all the stories of Christian martyrs. Is it possible that these were apocryphal stories of pagans who resisted the imposition of Christianity, stories that were then whitewashed in order to be assimilated into the Christian mythos?

Meanwhile, there was a third group of individuals during the time of transition: the pagan Platonists. There were two paths of Platonists: one which taught that one could approach God only by contemplating their own soul and knowing themselves; the other emphasized the beauty of the world as the means by which one might know God. These two ideas became the property of the educated man of the time, including Jews and early Christians. However, it was among the intellectual Jews of Alexandria that these ideas were given a subtle twist: a man could not know himself and thereby know God, he must give up any idea of ever knowing himself and resign himself to the 'grace' of god. God might choose a man and apply grace, but man must never think he could choose God and achieve grace. The Christian theologians took this idea and sculpted it to fit their new ideas of Christ and Redemption.

Many pagan ideas were adopted into Christian theology, but the chief difference was, as I have noted, the idea of sin being a personal matter, a personal fault, a sort of 'scapegoat principle' writ on the human soul. The pagans never considered it necessary to die with one's sins forgiven, and the dramatic deathbed scenes of Christianity, praying for the afterlife of the individual, were novel and rapidly spread. Pagans had prayed *to* the dead; Jews and Christians prayed *for* them. Fearing their own inevitable fault and sinful nature, Christians also prayed that the dead would intercede with God on their behalf. Christians, like pagans, continued the practice of feasting and celebrating death, with the added element of 'intercession' giving new meaning to the event.

Further along, there was another event in the pagan world of Europe that helped bring Christianity to dominance in the West of Europe, and brought another player onto the stage: Islam.

It was a warm, clear afternoon in the capital. The bustle of metropolitan commerce and tourism filled the streets. Small sailing vessels dotted the sheltered waters within sight of the government buildings, riding on a soft southerly breeze. The Sun sparkled on the gentle swells and wakes, lending a luminous glow to the poppies and tulips nodding in the parks along the water's edge. All was in order.

But suddenly, the sky brightened as if with a second, more brilliant sun. A second set of shadows appeared; at first long and faint, they shortened and sharpened rapidly. A strange hissing, humming sound seemed to come from everywhere at once. Thousands craned their necks and looked upwards, searching the sky for the new Sun. Above them a tremendous white fireball blossomed, like the unfolding of a vast paper flower, but now blindingly bright. For several seconds the fierce fireball dominated the sky, shaming the Sun. The sky burned white-hot, then slowly faded through yellow and orange to a glowering copper-red. The awful hissing ceased. The onlookers, blinded by the flash, burned by its searing heat, covered their eyes and cringed in terror. Occupants of offices and apartments rushed to their windows, searching the sky for the source of the brilliant flare that had lit their rooms. A great blanket of turbulent, coppery cloud filled half the sky overhead. For a dozen heartbeats the city was awestruck, numbed and silent.

Then, without warning, a tremendous blast smote the city, knocking pedestrians to the ground. Shuttered doors and windows blew out; fences, walls, and roofs groaned and cracked. A shock wave raced across the city and its waterways, knocking sailboats flat in the water. A hot, sulfurous wind like an open door into hell, the breath of a cosmic ironmaker's furnace, pressed downward from the sky, filled with the endless reverberation of invisible landslides. Then the hot breath slowed and paused; the normal breeze resumed with renewed vigor, and cool air blew across the city from the south. The sky overhead now faded to dark gray, then to a portentous black. A turbulent black cloud like a rumpled sheet seemed to descend from heaven. Fine black dust began to fall, slowly, gently, suspended and swirled by the breeze. For an hour or more the black dust fell, until, dissipated and dispersed by the breeze, the cloud faded from view.

Many thought it was the end of the world... (Lewis 1996)

The above quote is a reconstruction of events in Constantinople, 472 AD, extracted from *Rain of Iron and Ice* by John S. Lewis. According to Dr. Lewis, whose fanciful scenario of what it might be like to witness an overhead cometary fragment explosion, our earth actually experiences these types of events rather often, even if somewhat irregularly. Explosions in the sky – some of them enormous – have, according to him and many other scientists, profoundly affected the history of humanity. One obvious prospect is the great Antioch earthquake of 526 AD which was described by John Malalas (already quoted in a previous chapter):

> ... those caught in the earth beneath the buildings were incinerated and sparks of fire appeared out of the air and burned everyone they struck like lightning. The surface of the earth boiled and foundations of buildings were struck by thunderbolts thrown up by the earthquakes and were burned to ashes by fire ... It was a tremendous and incredible marvel with fire belching out rain, rain falling from tremendous furnaces, flames dissolving into showers ... As a result, Antioch became desolate ... In this terror up to 250,000 people perished. (Malalas, quoted in Baillie 2006, 130)

Strangely, historians, as a group, don't speak about such things. But the evidence is mounting.

The change of the Western world from pagan to monotheistic – Judaism, Christianity, Islam – effectively changed how people viewed themselves and their interactions with their reality. Today we live with the fruits of those changes: war without end. Constantine's victory paved the way for the recognition of Christianity by the Roman Empire and the union of Church and State that lasted for nearly 1,500 years and may, in fact, still be strange bedfellows though they have pulled up the covers to hide their relationship. An inscription quoting an ancient Hittite king informs us that a great prince needs the priests to instill the fear of the gods into the people so that they will do the will of the king, and the religion needs the protection of the ruler to impose its practice. So it has been for millennia.

Clube and Napier write:

[W]ithin these last few years, it has been found that there is a great swarm of cosmic debris circulating in a potentially dangerous orbit, exactly intersecting the Earth's orbit in June (and November) every few thousand years. More surprisingly, perhaps, it has been found that *the evidence for these facts was in the past deliberately concealed.* When the orbits exactly intersect, however, there is a greatly increased chance of penetrating the core of the swarm, a correspondingly enhanced flow of fireballs reaching the Earth, and a greatly raised perception that the end of the world is nigh. This perception is liable to arise at other times as well, whenever fresh debris is formed, but deep penetrations occurred during the fourth millennium BC, again during the first millennium BC, taking in at their close the time of Christ, and will likely take place yet again during the millennium to come. Christian religion began appropriately enough, therefore, with an apocalyptic vision of the past ... once the apparent danger had passed, truth was converted to mythology in the hands of a revisionist church, and such prior knowledge of the swarm as existed, which now comes to us through the works of Plato and others, was later systematically suppressed. ... the Christian vision of a permanent peace on Earth was by no means universally accepted, and it was to undergo several stages of "enlightenment" before it culminated with our present secular version of history, to which science itself subscribes, perceiving little or no danger from the sky. The lack of danger is an illusion, however, and the long arm of an early Christian delusion still has its effect. ...

The idea of a terrible sanction hanging over mankind is not, of course, new. Armageddon has been widely feared in the past, and it was a common belief that it would arrive with the present millennium ... Sometimes the proponents of such ideas escape to newfound new lands where in due course they meet opposition of a homegrown kind. In the United States, for example, despite freedom of speech, old traditions of cosmic catastrophe have recurred from time to time, even in the present century, only to be confronted by Pavlovian outrage from authorities. That being the case, it is perhaps ironic that elections in the United States are generally held in November, following the tradition of an ancient convocation of tribes at that time of the year, which probably had its roots in a real fear of world-end as the Earth coincided with the swarm.

In Europe, the millennium was finally dispensed with when an official "providential" view of the world was developed as a counter to ideas sustained during the Reformation. Indeed, to hold anything like a contrary view at this time became something of a heresy and those who were given to rabble-rousing for fear of the millennium were roundly condemned. To the extent that a cosmic winter and Armageddon have aspects in common, therefore, authoritarian outrage is nothing new.

... Enlightenment, of course, builds on the providential view and treats the cosmos as a harmless backdrop to human affairs, a view of the world which

61

Academe now often regards as its business to uphold and to which the counter-reformed Church and State are only too glad to subscribe. Indeed, it appears that repeated cosmic stress – supernatural illuminations – have been deliberately programmed out of Christian theology and modern science, arguably the two most influential contributions of western civilization to the control and well-being of humanity.

As a result, we have now come to think of global catastrophe, whether through nuclear war, ozone holes, the greenhouse effect of whatever, as a prospect originating purely with ourselves; and because of this, because we are faced with "authorities" who never look higher than the rooftops, the likely impact of the cosmos figures hardly at all in national plans. ...

... *A great illusion of cosmic security thus envelops mankind, one that the "establishment" of Church, State and Academe do nothing to disturb.* Persistence in such an illusion will do nothing to alleviate the next Dark Age when it arrives. But it is easily shattered: one simply has to look at the sky. The outrage, then, springs from a singularly myopic stance which may now place the human species a little higher than the ostrich, awaiting the fate of the dinosaur. [my emphasis] (Clube & Napier 1989, 11–13)

An abundance of fireballs and repeated comet sightings apparently excites a lot of 'eschatological activity' – predictions that the world is going to end – that can lead to all kinds of social unrest, which is, as Clube points out, highly undesirable to the ruling elites. After all, if people are thinking the world is going to end, they generally blame it on their rulers for being so corrupt and evil. The way they usually handle that sort of thing is to create an ostensible enemy who is responsible for it all, get a war going that soothes everyone's 'end of the world blues' and kills most of them in the bargain. A rather devious strategy indeed.

But it is here that we are going to discover how witches came to be associated with Halloween.

4. A Hundred Years of War, Witches, and the Inquisition

The so-called 'Hundred Years War' was a conflict between France and England, over claims by the English kings to the French throne. It was punctuated by several brief and two lengthy periods of peace before it finally ended in the expulsion of the English from France, with the exception of the Calais Pale. We notice that this state of conflict was already in motion about ten years before the Black Death fell on Europe.

When one studies the history of the Black Death and the Hundred Years War side by side, the thing that stands out is that whatever was going on then, there were conscienceless people taking advantage of the situation of confusion and terror. For example, we read the following:

> This would be a war of devastation. Villages and crops were burned, orchards were felled, livestock seized and residents harried. On Edward's entry into France he spent a week torching Cambrai and its environs. More than 1,000 villages were destroyed. France did what it could in England, at the war's onset seamen ventured to the southeastern coast of England to burn and ravage there. Much plunder was taken back to England and the thought of acquiring ill-gotten gain enticed many to support the war.
>
> Cruelty abounded. After the city of Limoges was captured and burned, Edward ordered the townsmen executed. Much of Artois, Brittany, Normandy, Gascony and other provinces were reduced to desolation (circa 1355 to 1375) and France did the same to the provinces that sided with England. Walled towns were safe during the early period of the war, but churches, monasteries, villages and rural areas were ruined.
>
> Truce and treaty were not observed. The "Free Companies" went into action, bandits of either English, French or hired mercenaries led by captains that dominated large areas and levied tribute on towns, villages and churches. They also seized women, took clergymen as accountants and correspondents, children for servants and plundered. (Cheney 1936)

Albert A. Nofi and James F. Dunnigan tell us:

For the first few years of the war there wasn't much happening except English raids into France and Flanders. Then, in the 1340s, England and France took opposite sides in the long-running civil war over who should be the duke of Brittany. In 1346 this resulted in a French invasion of Gascony and the shattering French defeat at Crecy. The English then rampaged through western France, until a truce was signed in 1354 (brought on by the devastation of the Plague, which hit France heavily in 1347–48).

The truce didn't last. In 1355, the war began again. In 1356 another major battle was fought at Poitiers and the French king was captured. English raids continued until 1360, when another truce was signed.[11]

One wonders if all this is not history written after the fact, placing the blame of cometary destruction and social unrest on a 'hundred years war'? As evidence to support this, it seems that the weather was going crazy. Clube and Napier write:

> One chronicler at least reports of the most immediate cause of the plague in 1345 that "between Cathay and Persia there rained a vast rain of fire; falling in flakes like snow and burning up mountains and plains and other lands, with men and women; and then arose vast masses of smoke; and whosoever beheld this died within the space of half a day ..." There seems little doubt also that *a worldwide cooling of the Earth* played a fundamental part in the process. The Arctic polar cap extended, changing the cyclonic pattern and leading to a series of disastrous harvests. These in turn led to *widespread famine, death and social disruption*. In England and Scotland, there is a pattern of abandoned villages and farms, soaring wheat prices and falling populations. In Eastern Europe there was *a series of winters of unparalleled severity* and depth of snow. The chronicles of monasteries in Poland and Russia tell of cannibalism, common graves overfilled with corpses, and migrations to the west. Even before the Black Death came, then, a human catastrophe of great proportions was under way in late medieval times. Indeed, the cold snap lasted well beyond the period of the ... plague. A number of such fluctuations are to be found in the historical record, and there is good evidence that these climatic stresses are connected not only with famine but also with times of great social unrest, wars, revolution and mass migrations. [my emphasis] (Clube & Napier 1989, 274)

When reading the above account, it is difficult not to think of the overall similarity to our own era. There are differences in detail and in scale, but the dynamics of a world gone mad, incredible cruelty running rampant, and global climate fluctuations are the same as we see before us now.

---

[11] http://www.hyw.com/books/history/Hundred_.htm

One naturally wonders why the masses of people would put up with such a state of affairs since it was they – and not the elite – who took the brunt of the horrors. The answer then is the same as it is now. The masses of ordinary people support their leaders in war because of propaganda. During wartime, Church and State generally form an alliance and patriotic statements are used in church sermons to support the ruling elite. The goal of the government is always to make the masses hate the enemy that the leaders wish to destroy (or at least to take their attention off their own depredations on the body social). In addition to the propaganda of Church and State, governments will offer increased wages and new opportunities to those who fight in the war (like US-sponsored mercenaries in Iraq and elsewhere). Criminals are often released from prison to fight. Then and now, people are promised lands, goods, benefits of all kinds, if they join the war effort. In some cases, what is offered to the common man is just to be left alone in their 'normal life' and not hounded or ridiculed. All of this has been how wars have been supported since time immemorial, and nothing has changed. The lures of power and goods make people who have no conscience, or who are low on the social totem pole, enthusiastic to join in killing other people just like themselves.

Calvinism was one of the developments that came out of this period. As Clube notes, the Protestant Reformation was partly due to the fact that the powers of the time, the Catholic Church, had built their control system based

A depiction of social upheaval and terror at the time of the 'black death' and the 'hundred years war'.

on the Aristotelian system of 'God is in his heaven and all will be right with the world if you are a good Christian.' Obviously, they didn't want to talk about a cosmos run amok over which their vaunted God had no control. And the fact that things were running amok and the Church couldn't do anything about it (not to mention the corruption of the Church that was evident to the masses) gave ammunition to the Reformers who then were able to attract many followers just as Christianity attracted Constantine at a time when the pagan gods did not seem to be able to help in the face of cometary bombardment.

The Protestants thus were able to use the situation to their advantage, suggesting that it was the 'End of Times' and that this was all part of the plan and people would be saved if they would only come over to the Protestant side.

Of course, once the Protestants had 'won their place', so to say, they too had to establish authority and adopt the Aristotelian view. '*Now*, God is in his heaven and all will be right and there won't be any more catastrophic disruptions as long as everybody goes to church, tithes, and obeys the appointed authorities.'

This brings us to the topic of witch persecutions. From the early decades of the fifteenth century until 1650, continental Europeans executed between *two and five hundred thousand witches* (according to conservative estimates), *more than 85 percent of them being women* (Ben-Yehuda 1985). People of the time, and even later, really did believe in the reality of witchcraft and evil demons. Men like Newton, Bacon, Boyle, Locke and Hobbes firmly believed in the reality of evil spirits and witches. As historian and religious scholar J. B. Russell wrote, "*Tens of thousands of [witch] trials continued throughout Europe generation after generation, while Leonardo painted, Palestrina composed and Shakespeare wrote*" (Russell 1980, 79).

Witchcraft and witches have existed throughout history, though in a context completely different from that which came to be understood during the crusade against witches. The Old Testament pretty much ignores the topic, except to report an encounter between King Saul and the witch of Endor, and to include a law: "*Thou shalt not suffer a witch to live.*" But other than that, in a way that seems to bizarrely contradict that law, stories of witches in the Bible are surprisingly neutral. There is no conceptualization or elaboration of witches, devils, or any kind of demonic world. The world of the Old Testament is, in fact, a world surprisingly devoid of anything truly spiritual.

In ancient Greece and Rome, *magic* was used to produce rain, prevent hail storms, drive away clouds, calm the winds, make the earth bear fruit, increase wealth, cure the sick, and so on. It could also be used against one's enemies to deprive them of those desirable effects. These beliefs were widespread in the ancient world, influenced by Pythagoras and his Northern European

Druidic training, and generally, 'good magic' was lawful and necessary, and 'bad magic' was condemned and punished. The state even supported those who could purportedly do good magic. It depended on perspective whether you were a 'good' magician or a 'bad' one. That's probably why the English condemned Joan of Arc for being a witch and France turned around and canonized her.

The Graeco-Roman religious universe – the supernatural world – was not divided into extreme good and extreme evil. It was occupied by every shade and combination of all qualities exactly as existed in human society. In this world, magic was simply an attempt to harness the power of the Unseen, while religion occupied itself with respect and gratitude to Nature and its representatives for results. In this way, prayers and spells could be easily combined.

The witch or sorcerer was a person who had a method – a technology – that could be used to harness and activate supernatural powers for her/himself or for others. She or he could 'control' the forces of nature (at least, that is what they believed, and who are we to say that the truly ancient shamanic technicians couldn't?).

So, two points are important here: 1) witchcraft/sorcery was a technology, and 2) there was a definite distinction between good magic and bad magic, and context was all-important.

After the disintegration of the Roman Empire and the rise of Judeo-Christianity, many missionaries, on finding that the pagans had their own spectrum of local deities and beliefs, often sought to convert them by the simple expedient of canonizing the local gods so that the native population could continue to worship them under the aegis of Christianity. They became 'Christian saints' complete with invented hagiographies (as I mentioned above, possibly most 'Christian martyrs' were actually pagans killed by the Church). The old temples were converted into churches so that the pagans would come to familiar places of worship to hear mass and pray to their 'saints' just like always. Magical practices were tolerated because it was felt that the people would give them up naturally over time once they had become truly Christian.

Official Church policy held that any belief in witchcraft was an illusion. In the famous, but mysterious, *Canon episcopi*, we find a few clues:

> Some wicked women, perverted by the devil, seduced by illusions and phantasms of demons, believe and profess themselves in the hours of night, to ride upon certain beastes with Diana, the goddess of pagans, and an innumerable multitude of women, and in the silence of the dead of night to traverse great spaces of earth, and to obey her commands as of their mistress, and to be summoned to her service on certain nights. But I wish it were they alone who perished in

their faithlessness and did not draw many with them into the destruction of infidelity. For an innumerable multitude, deceived by this false opinion, believe this to be true, and so believing, wander from the right faith and are invalued in the error of the pagans ...

Wherefore the priests throughout their churches should preach with all insistence ... that they know this to be false and, that such phantasms are imposed and sent by the malignant spirit ... who deludes them in dreams ... Who is there who is not led out of himself in dreams, seeing such in sleeping which he never sees [when] waking? ... And who is so stupid and foolish as to think that all these things, which are only done in spirit, happen in the body?

It is therefore to be proclaimed publicly to all that whoever believes such things ... has lost his faith. (Kors & Peters 1972, 29–31)

The origin of this document, which Kors and Peters date to 1140, is not clear. It has been attributed to an obscure meeting, the Council of Anquira, held possibly in the 4th century. Although there is no record of this council, the statement on witchcraft was adopted by later canonists as official policy. What it does tell us is that there were, apparently, worshippers of the pagan goddess Diana who had profound experiences that were declared to be delusions brought on by the Devil. Right here we see how the Goddess was replaced by Satan the deluder. It is interesting to compare the description of what these ancient witches were said to be doing with the activities of ancient Siberian shamans. One is also reminded of the Paleolithic cave paintings when reading, *"in the hours of night, to ride upon certain beastes with Diana, the goddess of pagans ..."* This is a precious clue to the fact that the Paleolithic religion and its shamanic lines did survive for thousands of years.

In any event, for more than six centuries this was the official attitude of the Church toward witches – that it was an illusion or delusion or just the product of dreams, and *"whoever was 'so stupid and foolish' as to believe such fantastic tales was an infidel."* That, apparently, applied to monks and priests, and the general public as well. The important point here being that you had to believe in witches to persecute them, and believing in them as real was against Church doctrine.

Taking into account the Black Death and the wars of the time killing off so much of the male population, one might suppose that there was an increase in unmarried women or women who had inherited estates when all other family members had perished. In short, women were becoming autonomous as a consequence. And certainly, women who had 'gifts' would be more likely to survive such calamities than those who did not.

The details of exactly what happened may be forever lost to us thanks to the cover-up of history instituted by Joseph Justus Scaliger in the sixteenth century as has been suggested by Clube and revealed in some detail by mathematician Anatoly Fomenko (though Fomenko does not take catastrophic destruction of society into account). The best we can do is to speculate.

The most spectacular 'witch' was Joan of Arc who was tried, condemned, and burned in 1431. Her trial and execution can clearly be seen as political, often a major underlying motivation for such accusations.

It could be said that the witch persecutions were simply a reviving of the Inquisition that had created similar rules for dealing with the Cathars two hundred years earlier. In order to fully understand how easily this attack on 'witches and sorcerers' could manifest legally and socially, we need to take a quick look at the beginnings of the Inquisition.

Many people think of the Inquisition as something that was started to eliminate witches and Devil worship, and the word conjures images of the rack and iron maidens and all kinds of bizarre and twisted instruments of torture. Sure, torture was a big part of the Inquisition, but not as much as some people might think. You have to remember that the Inquisition began during a period of history when human life was treated so casually by some that cutting off noses or ears or hands, or gouging out eyes was not unheard of as a legal punishment.

After years of brutal massacres, of destruction of the land, of some of the most horrible events ever to bear witness of man's inhumanity to man, Pope Gregory IX decided that it was only results that counted. He intended to wipe Catharism from the face of the earth. He must have sat up at night to create the bizarre system that was put into place to deal with heresy. Or maybe such thinking just comes natural to psychopaths.

First, he created special papal legates who were granted wide powers of prosecution similar to what we have today in the Homeland Security nonsense in the USA, and MIVILUDES in France, and sent them out all over Europe. The men chosen for this task were clearly psychopaths, and their mission was to spread terror all over Europe.

A 'witch' being burned at the stake.

Gregory staffed the Episcopal palaces of the South of France with psychopathic bishops who offered a cash bounty to anyone who betrayed a heretic. The inducements to betray one's neighbor were surely tempting in the best of times. But in a time when starvation and destruction was everywhere after more than 20 years of the rampaging of the crusading armies, it was well-nigh impossible to resist. *The terms were that the property confiscated from the heretic was divided between the informer, the Church and the Crown.* Naturally, in a land that was financially devastated, where people were displaced and starving after years of being battered by this same Church and Crown, there were a lot of individuals offering up their neighbors for blood money. Maybe that is why the PTB have engineered the current economic crises: to soften people up to be more willing to turn against each other for a crumb? It's certainly easier to enforce full-spectrum dominance when the population cooperates.

Robert le Bourgre, which means 'the bugger' (suggesting the contempt in which he was held by the people), terrorized formerly peaceful northern France. Another legate, Conrad of Marburg found unsuspected heretics everywhere in the Rhineland. Thousands were sent to the stake, often on the same day that they were accused. Conrad rode about on his mule with two assistants, bringing terror to every village and town they approached. Apparently, even the regular clergy saw through this nonsense and finally decided to do something about it. On 30 July, 1233, a Franciscan friar, driven to act in the name of justice, intercepted Conrad and murdered him. That was the living end for the Pope. He was *not* amused and he was not going to be denied his imperialist ambitions. He turned to the Dominicans.

In the spring of 1233, papal inquisitors were appointed in Toulouse, Albi and Carcassonne. These inquisitors were succeeded in an unbroken line for 600 years. (I would suggest that they have now been replaced by France's 'anticult' organization, MIVILUDES.) Thousands of people were summoned to testify before inquisitors. The questions were repetitive, designed to plant doubt in the mind of the person being interrogated as to what exactly the inquisitor knew, and who had told him. A person suspected of Cathar sympathies was not always informed of the charges hanging over his head; if apprised of the danger, he had no right to know who his accusers were; and if he dared to seek legal help, his lawyer could be charged with abetting heresy. Whatever the verdict of the inquisitor – who was prosecutor, judge and jury – no appeal was allowed. Anyone could be held indefinitely in prison for further questioning without cause of explanation. Nowadays, we call them 'enemy combatants'.

The inquisition destroyed the bonds of trust that hold societies together. Informing on one's neighbor became not only a duty, but a necessary survival strategy. For 100 years, the Inquisition was a fact on the ground of life in the Languedoc. The arrival of an inquisitor in a town was the occasion for demeaning displays of moral collapse just as the arrival of AIPAC or MIVILUDES is the occasion for legislators and journalists to take off their shirts and display the wide yellow streaks down their backs.

In theory, of course, no one could be punished if no one talked because the inquisitor could not act without a writ of denunciation, but in practice, no community possessed the cohesion needed to combat the power of a secret tribunal. And that is the problem that the masses of humanity face today. Because pathological deviants exist among us, and we do not have the accurate psychological knowledge to identify them and weed them out, we are constantly at risk of being betrayed by the 'right-wing authoritarians' that lurk in our society just waiting to turn *us* in because they have a grudge against us or want our possessions.

They don't even have to be genetic deviants to be dangerous. In America today, the masses have been thoroughly conditioned by watching reality TV and *Survivor*, they have learned the rules of the psychopath and live by them: *Do unto others before they do unto you*. The same is true in France due to their psychologically traumatic education/indoctrination system. And so it was in the Languedoc, the historical model for what is happening in France and the United States today, and for what happened in Germany under Hitler and many other examples from history, was created, tested and refined.

Upon his arrival in a town, the inquisitor consulted with the local clergy. All males over the age of 14, and females over the age of 12, were required to make a profession of faith in the Catholic Church. Those who didn't were the first to be questioned.

Then, the inquisitor would give a speech in which he invited the people to spend some days thinking very, very hard about their activities past, present and future, and to come forward in the following week to give confidential depositions. After a seven-day grace period, those who had not denounced themselves would be issued a summons. The punishments ranged from loss of property to loss of life. As you might guess, everyone was outdoing themselves in an effort to denounce themselves and each other.

Aside from the capital crime of being a Cathar, punishable offenses included sheltering a Cathar, or even failing to report any instance of heresy. The real proof of genuine piety toward the Catholic Church was defined as the number of people the sinner was willing to betray.

It only took ten years for the Inquisition, the work of a few psychopathic fanatics, to become a proficient bureaucracy that lasted for the already mentioned 600 years. It employed hundreds of individuals who interrogated thousands of people with such monotonous regularity that a regular 'glossary' was established for the 'workers'.

Armed with a list of proposed offenses to be considered 'heretical' or 'supporting heretics', which included just knowing that a heretic had crossed one's property and failing to report it, the Inquisition proceeded to intimidate the population of Europe on a scale that was impossible to imagine. The sheer numbers of people called to testify, and re-called to testify again and again, was staggering. In a strange twist of historical irony, the Cathars – who believed that the material world was evil and irrelevant – inspired the codification of the police state: the ultimate material theology.

A cross-referenced compendium of the confessions extracted from tens of thousands of people was compiled, creating a map of the mental landscape of the Languedoc. The more than five thousand transcripts of interrogations that survive represent only a small fraction of the work of the Inquisition. Inquisitors' manuals were created to serve as guides for the growing number of papal courts in Europe. These manuals reminded the inquisitors that they were in the business of saving souls, but I think that the distinction was lost on those whose lives were lost or ruined by the judgments of the Inquisition.

Languedoc was, essentially, the laboratory for repression.

The reputation of the Inquisition was enhanced by the talented Inquisitor of Toulouse, Bernard Gui, who was the villain in Umberto Eco's *The Name of the Rose*. The Inquisitors persuaded a handful of captured Cathars to convert and sell their testimony. Sicard de Lunel of Albi gave the friars an exhaustive list of Cathar sympathizers, even fingering his own parents. Anyone who had ever helped him in his life as a Cathar, whether they had just given him a bed for the night, a bit of food, or even a jar of honey, were hauled in to be punished – just on his word. He and several others like him were lodged thereafter in a castle outside of Toulouse in the medieval version of the 'witness protection program'. Sicard was well paid for his perfidy and lived to a ripe old age.

The use of torture was delicately referred to as 'putting the question'. In the Languedoc, successive waves of highly trained inquisitors, aided by informers and torturers, fired by the totalitarian creed of the Catholic Church, with detailed manuals and expanding registers of 'intelligence', slowly but surely ground Catharism – and a marvelous, productive society – into oblivion. Thousands of dramas of conscience ended in the dungeons or in fires quenched with blood. By the end of the century, only the truly heroic dared to say that this world was evil …

It was not a legal system; it was a system designed to create fear. This 250-year-old Catholic system of terror was the system that was handily available at the beginning of the witch persecutions, though, curiously, the first persecutions were not ecclesiastical but rather *political*.

In 1397, a man named Stedelen was accused of being a witch in Simmental, Switzerland, after the harvest had failed at his village. According to his accusers, Stedelen used black magic to destroy the crops by allegedly sacrificing a black rooster on the Sabbath at a crossroad and placing a lizard under the doorway of a local church. Peter von Greyerz, the judge, was a firm believer in witchcraft, which he believed had been introduced in Simmental by a noble man called Scavius in 1375. He was killed by his many enemies, but he had a student, who according to Greyerz had been the tutor of Stedelen.

There was no real evidence presented, of course, but Stedelen had allegedly become an expert on magic and supposedly learned to steal manure, hay and such from others' fields by magic, create hail and thunderstorms, make people and animals sterile, make horses crazy when he touched their hooves, fly and terrify those who captured him. Greyerz also accused Stedelen of having taken the milk from the cows of a married couple in order to make the wife miscarry. After torture, Stedelen confessed to summoning forth demons as part of a pact with the Devil. His trial took place in a secular court following which he was burned at the stake.

Greyerz believed that there was a Satanic cult, whose members swore themselves to the Devil and ate children at the churches at night. He continued his persecutions and once tortured a woman to confirm his beliefs.

In 1415–1419, in the Duchy of Savoy, there was a civil war between clans of the nobility. Various noble families had rebelled against the Raron family, and the masses were drawn into the conflict. This had gone on for some time and, by 1428, the entire society in that area was in a state of great tension. No one knows who came up with the idea that all the troubles were due to witches, but on 7 August 1428, delegates from seven districts in Valais demanded that the authorities investigate alleged, unknown witches and sorcerers. Anyone denounced as a sorcerer by more than three people was to be arrested. If they confessed, they were to be burned at the stake as heretics, and if they did not confess, they would be tortured until they did so. Also, those pointed out by more than two of the judged sorcerers were to be arrested.

These accusations, trials and executions were probably seen by other elite individuals – or individuals who wished to become members of the elite by seizing the property of those they resented or envied – as a handy way to deal with many problems. The craze rapidly spread north into Germany, then to France and Switzerland. The main accusations consisted of:

- flying through the air and plundering wine cellars;
- lycanthropy – to have killed cattle in the shapes of werewolves;
- to have made themselves invisible with herbs;
- to have cured sickness and paralysis caused by sorcery by giving it to someone else;
- to have abducted and eaten children;
- to have met Satan and learned magic from him;
- to have planned to deprive Christianity of its power over humanity.

From this list, we can surmise the kinds of troubles that the people were suffering: starvation leading to theft and vandalism, destruction of their livestock, widespread sickness, loss of children, and, undoubtedly, even cannibalism. It was clearly a very difficult time.

One hundred years after the Black Death had destroyed about half of Europe's population, the Hundred Years War was coming to an end, things were still extremely difficult, and in order to restore order and control on the recovering society someone had to be blamed (definitely *not* cometary explosions). The problem was, of course, how to get around the *Canon Episcopi*. It was necessary to diminish this official Church policy in order to even have a 'witch craze'. So, the first attacks were made on the validity of the document itself.

In 1450, Jean Vineti, inquisitor at Carcassonne, identified witchcraft with heresy, and in 1458, Nicholas Jacquier, inquisitor in France and Bohemia, identified it as a *new* form of heresy – that is, *contemporary witches were claimed to be different from the ones that the document was about*. In 1460, Visconti Girolamo, inquisitor, professor, provincial of Lombardy, stated that the act of defending witchcraft (or witches) was itself heresy.

Incremental steps were being made toward establishing an official standard, but still, when Kramer and Sprenger (both members of the Dominican Order and inquisitors for the Catholic Church) wrote the *Malleus Maleficarum* and submitted it to the University of Cologne's Faculty of Theology on 9 May 1487, seeking its endorsement, it was roundly condemned as unethical and illegal. The Catholic Church banned the book in 1490, placing it on the *Index Librorum Prohibitorum*, and Kramer was denounced by the Inquisition. It should be noted here that, in 1484, Kramer had attempted a systematic persecution of witches in the region of Tyrol, which bombed dramatically. Kramer was thrown out of the territory and dismissed by the local bishop as a 'senile old man'. According to historian of the Church Diarmaid MacCulloch, writing the book was Kramer's act of self-justification and revenge.

The main thrust of the *Malleus* was to systematically refute the *Canon Episcopi* and to discredit those who expressed skepticism about the reality of witchcraft, to claim that witches were more often women than men, and to educate prosecutors on how to expose and convict them. Experts say that the *Malleus* was based, in part, on the *Formicarius* by Johannes Nider, written about ten years earlier. Before Nider, magic was thought to be performed by educated men who performed intricate rituals.

In Nider's *Formicarius*, the witch is described as illiterate and female. Unfortunately, Johannes Gutenberg's printing press – a product of the Renaissance – allowed the work to spread rapidly throughout Europe. This crystallization is what resulted in the beginning of the witch craze itself. The idea that any person could harm another via magic simply by devoting themselves to the worship of Satan – especially women, who had been long viewed as helpless and somewhat less than human – was terrifying and shocking.

As the craze spread over Europe, literally hundreds of thousands of women were burned at the stake. Children and even whole families were sent to be burned. The historical sources are full of horrifying descriptions of the tortures these poor people were subjected to. Entire villages were exterminated. One account says that all of Germany was covered with stakes and Germans were entirely occupied with building bonfires to burn the victims. One inquisitor is reported to have said: *"I wish [the witches] had but one body, so that we could burn them all at once, in one fire!"* (Trevor-Roper 1967, 152)

In the 1580s, the Catholic Counter-Reformation became dedicated witch-hunters also, going after Protestants, mainly. In France, most witches happened to be Huguenot. In Protestant areas, most witches were Catholic. It could be said that most cases of witch burnings were either personal or political or both. One victim was a judge who was burned in 1628 for showing 'suspicious leniency'. As the craze spread, the viciousness and barbarity of the attacks increased. The aforementioned judge, a Dr. Haan, confessed under torture to having seen five burgomasters of Bamberg at the witches' Sabbath, and they too were executed. One of them, a Johannes Julius, confessed under torture that he had renounced God, given himself to the Devil, and seen twenty-seven of his colleagues at the Sabbath. But afterward, from prison, he contrived to smuggle a letter out to his daughter, Veronica, giving a full account of his trial. He wrote:

> Now, my dearest child, you have here all my acts and confessions, for which I must die. It is all falsehood and invention, so help me God … They never cease to torture until one says something. If God sends no means of bringing the truth to light, our whole kindred will be burnt. (Trevor-Roper 1967, 157)

Protestants and Catholics accused each other, and the early decades of the 1600s were infected by a veritable epidemic of demons. This lasted until the end of the Thirty Years War. It is said that if the publication of the *Malleus Maleficarum* was the beginning of the terror, the Peace of Westphalia in 1648 was the end.

During this period, the distinction between good and bad magic vanished and witchcraft became something purely evil and almost totally female. The pluralistic conception of the supernatural world also vanished and we were left with only a very good God who was, however, seemingly impotent in the face of evil mankind in cahoots with a very evil Devil. Well, not exactly 'mankind'; mostly 'woman-kind'.

In recent times, the *Malleus* has been examined critically, though not by individuals with any awareness of the cosmic events of the time. Nevertheless, what they have observed has a bearing on our subject here. In his article, 'Sexy Devils', Dale Keiger writes:

One evening 10 years ago, Walter Stephens was reading *Malleus Malificarum*. The *Malleus*, as scholars refer to it, would not be everyone's choice for a late-night book. Usually translated as "The Hammer of Witches", it was first published in Germany in 1487 as a handbook for witch hunters during the Inquisition. It is a chilling text – used for 300 years, well into the Age of Reason – that justifies and details the identification, apprehension, interrogation, and execution of people accused of consorting with demons, signing pacts with the devil, and performing maleficia, or harmful magic.

"It was 11 at night", Stephens recalls. "My wife had gone to bed, and on the first page (of the Malleus) was this weird sentence about people who don't believe in witches and don't believe in demons: 'Therefore those err who say that there is no such thing as witchcraft, but that it is purely imaginary, even although they do not believe that devils exist except in the imagination of the ignorant and vulgar, and the natural accidents which happen to man he wrongly attributes to some supposed devil.'"

That convoluted sentence dovetailed with a curious line Stephens knew from *Il messaggiero*, a work from 1582 by the Italian poet Torquato Tasso: "If magicians and witches and the possessed exist, demons exist; but it cannot be doubted that in every age specimens of the former three have been found; thus it is unreasonable to doubt that demons are found in nature."

Stephens, the Charles S. Singleton Professor of Italian Studies in the Hopkins Department of Romance Languages, is a literary critic, and he sensed that something intriguing was going on beneath the text on the page. Tasso, and especially the Malleus' author, a Dominican theologian and Inquisitor named Heinrich Kramer, had in their works invested a striking amount of energy in refuting doubt about the existence of demons. What was that about?

For the next eight years Stephens read every treatise he could find on witchcraft, as well as accounts of interrogations, theological tracts, and other works (his bibliography lists 154 primary and more than 200 secondary sources). Most of the 86 witchcraft treatises he cites had been written in Western Europe in the 15th, 16th, and 17th centuries, and one after another (including the Malleus) contain accounts of sexual intercourse with Satanic spirits. Why? Were the authors' remorseless misogynists hell-bent on portraying

The inquisition mercilessly tortured thousands of innocent people.

women in the worst possible light? Were they lurid, repressed celibates who got off by writing accounts of demon sex? Stephens didn't think so; the texts, in his view, didn't support that reading. Elsewhere in the Malleus he had found a key reference to accused witches under torture as being "expert witnesses to the reality of carnal interaction between humans and demons". These guys are trying to construct proofs that demons exist, he thought. They're trying to convince skeptics. And then he thought *they're trying to convince themselves.*

Stephens' thesis profoundly revises the conventional wisdom about centuries of cruelty and injustice. The great European witch hunts, he says, were *the outgrowth of a severe crisis of faith.* The men who wrote books like the Malleus, men who endorsed the torture and burning of tens of thousands of innocent people, *desperately needed to believe in witches, because if witches were real, then demons were real, and if demons were real, then God was real. Not just real but present and attentive.* Carefully read the works composed by the witchcraft authors, Stephens says, and you will see *how profoundly disturbed these educated, literate men were by their accumulating suspicions that if God existed at all, He wasn't paying much attention to the descendants of Adam.*

… The Church itself fractured, riven by massive organized heresies, and by a schism that led to as many as three men simultaneously laying claim to be the true pope. How could a world created by a watchful, benevolent, and engaged God be such a mess? (*Johns Hopkins Magazine* 2002; emphases mine)

The fourteenth and fifteenth centuries were a time when the forces of nature ran amok, and the powers of the time needed to save face and keep control: how was it that they, who overcame paganism by the promise that their God could protect his believers from the forces of nature, now were exposed as totally incompetent to do so? There was, undoubtedly, a resurgence of paganism and to prevent that it was probably seen as ideal to blame the enemy – pagans – for destruction that they had nothing to do with.

Protestantism was on the rise, of course, but its proponents did not see it as politic to go after the Mother Church, which still held a great deal of power, so *some other sin-bearer had to be found.* At the end of the Hundred Years War and the Black Death, and the Thirty Years War – all of which may have been periods of cometary destruction on the earth – the witch persecutions were utilized to hush up completely any hint that the earth was not securely hung in space, and history and truth was suppressed with blood and burning human flesh.

Such persecutions were a means of controlling those who uttered 'heresies' against the 'providential' order of the universe established by the Church and State, like pointing out that an increased number of fireballs and comet sightings may very well suggest that the planet and its inhabitants were in potential danger. This was the period of Galileo, after all, and he was accused of being a heretic for not supporting the potency of God Almighty.

Another point to note is that, before this period, witches were still comprehended as beings that could use a technology to control the powers of nature – shamanic. After this period, they were known as beings that *only* channeled evil into the world because they were under the control of the Evil One. They were all purely Satan's puppets and no good could ever come from them. The *Malleus Maleficarum* specifically mentions that *"witchcraft is chiefly found in women because they are more credulous and have poor memories"*, and because *"witchcraft comes from carnal lust, which is in women insatiable"* (Sprenger & Kramer 1968, 41–48).

The political uses of these ideas should be obvious. Sprenger and Kramer, *et al.*, came along and wrote books describing healthy, competent, intelligent women as witches, and the problem was solved. All the excess women (or anybody, for that matter) can be gotten rid of; all the autonomous women with property can be done away with and their property confiscated; and, at the same time, the psychological control of men over women, re-establishing the subservience of women to the Church, can be accomplished in one fell swoop. (One also has to consider the destruction of many genetic lines of powerful women – shamanic lines – in this process, which has been ongoing, so it seems.)

One of the most distressing results of this change in attitude toward witches was the creation of witchcraft as a systematic *anti-religion in the minds of its persecutors*; it became the opposite of everything that Christianity – both Catholic and Protestant – stood for. *Witchcraft as an elaborated system of religion was unknown before the fifteenth century* (this is why modern-day reconstructions are not likely to be very accurate). *This was a period in which a theory of supernatural demons was invented and crystallized as an explanation for the evils that fell upon mankind.* How else to explain the Black Death which killed indiscriminately in spite of the prayers and supplications of the priests of the Christian Church, both Catholic and Protestant?

It seems that the legends of gods fighting in the skies (the break-up of a giant comet 13,000 years ago) were later corrupted into certain Gnostic ideas such as the 'cosmic error'. Certainly, at a certain level, there is duality, otherwise nothing would exist, but this Gnostic take on things went way too far with these ideas (see *The Other God* by Stoyanov for a better understanding of Gnosticism, keeping in mind the work of Victor Clube and Bill Napier).

The 'witch myth' was created in the late 1400s in reaction to the Black Death – cometary destruction on an almost unimaginable scale – and this myth consisted of a whole, coherent system of beliefs, assumptions, rituals, and 'sacred texts' that had never existed until this time and that were created by a couple of psychopathic accusers. The Dominicans developed and popularized the conceptions of demonology and witchcraft as a negative image of the so-called 'true faith', and the Protestants were just as busy.

What all this means is that being a 'witch' in the time of the witch persecutions must have meant something more akin to following a dualist belief system similar to the Cathars, being an observer of nature, astronomy, speaking truth to power; really, more like how Burton Mack described the early Jesus People. It probably also meant being able to 'see the unseen' in terms of cosmic, social and human energies, to 'walk between worlds' as the Paleolithic shamans did, and to use these abilities on behalf of other people. Perhaps the image of the witch flying on her broom across the face of the full moon was actually a corrupted ancient symbol of a comet with a tail personified as woman?

A comet came and nearly destroyed humanity at the end of October 13,000 years ago, and impacting debris from the same comet brought Judaism, Christianity, Islam and, later, the imposition of Christianity on the Western world. Later still, the same comet stream brought the Black Death and the persecution of witches, both male and female. This scapegoating was utilized to get rid of a lot of individuals who threatened the status quo – the control over the masses – and that included a great many strong, independent women. And so, today, we associate witches with Halloween, the end of October, and the anniversary of the destruction of nearly all life on earth. It is just a variation on the 'Eve ate the apple and brought about the fall in Eden' story, created by psychopaths who hate women and all they stand for: Creation, Nurturing and Service to Others.

Indeed. The calamities of that time – of *any* time – assault religious faith. And anyone who talks about such calamities in a reasonable and factual way as just what Nature does, and who backs it up with scientific data, *must* be silenced because they threaten the very foundation of Western civilization: Judeo-Christianity, uniformitarianism, and fascist control of humanity.

5. Thirty Years of Cults and Comets

In previous chapters we have looked at how the Black Death was probably a period of cometary fragment bombardment leading to mass death on an unimaginable scale. In today's world, the equivalent would be the deaths of two, possibly three billion people planet-wide and many animals as well. Just contemplating what humanity would do with that many bodies to be disposed of is daunting, not to mention considering how society would continue. The Black Death was no respecter of rank, either: the elites died in proportionate numbers to the masses of ordinary people.

In our discussion of the Hundred Years War, we learned that a great cover-up was effected at the end of it all and this was mainly to reestablish the religious control of the masses because, of course, religious control has always been the right arm of princes and governments.

Inasmuch as *it was popularly believed that the continued sterility of many years was caused by witches through the malice of the Devil*, the whole country rose to exterminate the witches. *This movement was promoted by many in office, who hoped wealth from the persecution.* And so, from court to court throughout the towns and villages of all the diocese, scurried special accusers, inquisitors, notaries, jurors, judges, constables, dragging to trial and torture human beings of both sexes and burning them in great numbers. Scarcely any of those who were accused escaped punishment. Nor were there spared even the leading men in the city of Trier. For the Judge, with two Burgomasters, several Councilors and Associate Judges, canons of sundry collegiate churches, parish priests, rural deans, were swept away in this ruin. So far, at length, did the madness of the furious populace and of the courts go in this thirst for blood and booty that there was scarcely anybody who was not smirched by some suspicion of this crime.

Meanwhile *notaries, copyists, and innkeepers grew rich.* The executioner rode a blooded horse, like a noble of the court, and went clad in gold and silver; his wife vied with noble dames in the richness of her array. The children of those convicted and punished were sent into exile; their goods were confiscated; plowman and vintner failed hence came sterility. *A direr pestilence or a more ruthless invader could hardly have ravaged the territory of Trier than this inquisition and persecution without bounds*: many were the reasons for doubting that all were really guilty. This persecution lasted for several years; and some of those who presided over the administration of justice gloried in the multitude of the stakes, at each of which a human being had been given to the flames. At last, though the flames were still unsated, *the people grew impoverished, rules were made and enforced restricting the fees and costs of examinations and examiners, and suddenly, as when in war funds fail, the zeal of the persecutors died out.*[12] [emphasis added]

Indeed, the question that led to the persecution of witches was a religious one: How could a world created by a watchful, benevolent, and engaged God be such a mess? Answering this question led to a growth industry in persons and institutions dealing death and destruction. We see a lot of that going on in our world today: the 'security industry' is booming in the mythical 'War on Terror'.

The Reformation divided Europe between Protestant regions and those loyal to the Pope, but Protestants took the crime of witchcraft no less seriously – and arguably even more so – than Catholics. Germany, rife with sectarian strife, saw Europe's greatest execution rates of witches – higher than those in the rest of the Continent combined. Witch hysteria swept France in 1571 after Trois-Echelles, a defendant accused of witchcraft from the court of Charles IX, announced to the court that he had over 100,000 fellow witches roaming the country. Judges responding to the ensuing panic by eliminating for those accused of witchcraft most of the protections that other defendants enjoyed. Jean Bodin in his 1580 book, *On the Demon-Mania of Sorcerers*, opened the door to use of testimony by children against parents, entrapment, and instruments of torture.[13]

The problem is, of course, that the primary targets in any such persecutions are those who talk about the calamities themselves and point out that the religious faiths are obvious failures and perhaps it might be better to look at the world rationally and scientifically. Such individuals must be accused of being witches or 'cults' and silenced because they threaten the very foundation of Western civilization, uniformitarianism and the fascist control of humanity by such elements.

---

[12] Linden's Gesta Trevirorum: http://www.sacred-texts.com/pag/twp/twp05.htm
(See this website for many first hand accounts and details of the witch persecutions: http://history.hanover.edu/early/wh.html)

[13] 'A Brief History of Witchcraft Persecutions before Salem': http://www.law.umkc.edu/faculty/projects/ftrials/salem/witchhistory.html

In Victor Clube's narrative report funded by the USAF and Oxford, the next important period of cometary calamity was the Thirty Years War. Let's look at a short timeline just to orient ourselves.

1337–1453: Hundred Years War.

1347/48–1351: Black Death (included in the time period of the Hundred Years War).

1400: Renaissance (begins as the Hundred Years War is ending).

1431: Joan of Arc burned at the stake for being a witch (included in the time period of the Hundred Years War).

1484: Pope Innocent VIII announced that Satanists in Germany were meeting with demons, casting spells that destroyed crops, and aborting infants.

1486: *Malleus Maleficarum* published.

1500: Witch persecutions begin.

1515: Outbreaks of witchcraft hysteria, subsequent mass executions begin.

1591: King James authorizes the torture of suspected witches in Scotland.

1600: Renaissance ends 'officially'.

1606: Shakespeare's *Macbeth* performed.

1616: **Thirty Years War begins.**

1642: Beginning of the English Civil War.

1643: The largest witch-hunt in French history occurred. For two years there were at least 650 arrests in Languedoc alone. The same time was one of intense witch-hunting in England, as the English Civil War created an atmosphere of unrest that fueled the hunting, especially under Matthew Hopkins.

1648: **Thirty Years War ends.**

1651: End of the English Civil War.

1660: Witch persecutions end. Europe saw between 50,000 and 80,000 suspected witches executed. About 80% of those killed were women. Execution rates varied greatly by country, from a high of about 26,000 in Germany to about 10,000 in France, 1,000 in England, and only four in Ireland. The lower death tolls in England and Ireland owe in part to better procedural safeguards in those countries for defendants.[14]

1682: England executes its last witch, Temperance Lloyd, a senile woman from Bideford. Lord Chief Justice Sir Francis North, a passionate critic of witchcraft trials, investigated the Lloyd case and denounced it as a farce. Witch-hunting shifted from one side of the Atlantic to the other, with the outbreak of hysteria in Salem in 1692.

I'm not too sure why the Renaissance is said to end in 1600. It looks to me more like it was probably the Thirty Years War that ended it. But, never mind, that's the date range agreed on by most scholars.

---

[14] http://www.law.umkc.edu/faculty/projects/ftrials/salem/witchhistory.html

The Thirty Years War was fought between 1618 and 1648, principally on the territory of today's Germany, and involved most of the major European powers. It began as an ostensible religious conflict between Protestants and Catholics and gradually developed into a general war involving much of Europe, related to the France-Habsburg rivalry for pre-eminence in Europe, which led later to direct war between France and Spain. (Notes to ponder: The Thirty Years War also pretty much spanned the reign of Louis XIII of France [1610–1643]. Galileo lived from 1564 to 1642. Many adherents of Catharism, fleeing a papal inquisition launched against their alleged heresies in France, had migrated into Germany and the Savoy. This may have been at the root of the initial religious conflict. In fact, Catharism may have fed the Protestant Reformation.)

The Thirty Years War was one which utilized mercenary armies to a great extent, and these hired killers were said to have devastated entire regions leaving the inhabitants to suffer widespread famine and disease which decimated the population. This affected primarily the German states and, to a lesser extent, the Low Countries and Italy. At the same time, it bankrupted many of the governmental powers involved. Again, the parallels to the situation on the planet today are striking.

The English Civil War, which began after the Thirty Years War had been going on for about 25 years (and was running out of steam and people), consisted of a series of armed conflicts and political machinations that took place between Parliamentarians (known as Roundheads) and Royalists (known as Cavaliers).

The question is, do we find any mentions of comets or other strange astronomical phenomena during this period of time? As it happens, we do.

David Herlicius published in 1619 a discourse on a comet that had appeared shortly before, in 1618, and enumerated the calamities that this comet, and comets in general, bring with them or presage:

> Desiccation of the crops and barrenness, pestilence, great stormy winds, great inundations, shipwrecks, defeat of armies or destruction of kingdoms … decease of great potentates and scholars, schisms and rifts in religion, etc. The portents of comets are threefold – in part natural, in part political, and in part theological.[15]

The seventeenth century was witness to numerous comet sightings, including those of 1618, 1664, 1665, and 1677. Inquiries into these comets produced a noteworthy number of scientific texts including Samuel Danforth's *An Astronomical Description of the Late Comet* (1665), John Gadbury's trea-

---

[15] 'William Whiston and the Deluge': http://www.varchive.org/itb/ecwhist.htm

tise *De Cometis* (1665), and Robert Hooke's 1678 report to the Royal Society, *Cometa*. These accounts[16] complemented the earlier work of Brahe and Kepler and helped to expand the emerging technical understanding of this particular cosmic phenomenon.

Another reference: *"This year (1618) brought on three bright comets."*[17]

Regarding Kepler: His observations on the three comets of 1618 were published in *De Cometis*, contemporaneously with the *Harmonice Mundi* (Augsburg, 1619).

My search for direct source material giving evidence of unusual events from this time has been rather frustrating. I have found that the only people reading the original documents are scholars who generally refer to the descriptions of the time as being hyperbole, or more or less 'religious' metaphor, so it is frustrating to find that these actual passages are quoted in the original language – generally German. So, I sent the material off to a German friend of mine and he quickly returned a translation.

In the journal, *German Life and Letters* 54:2, Geoffrey Mortimer published an article entitled 'Style and Fictionalisation in Eyewitness Personal Accounts of the Thirty Years War'. He writes:

> Eyewitness personal accounts of the Thirty Years War are of interest not only for their overt content, but as examples of how the process of writing itself can shape both the resultant text and the meaning derivable from it by the reader. Techniques adopted, probably unconsciously, by writers seeking to give force and point to their narratives, here collectively termed 'fictionalisation', add to well-known problems of eyewitness testimony to affect the historical evaluation of such sources.

As we will see, it is apparent that Mr. Mortimer hasn't been reading the work of Victor Clube. He goes on for some pages explaining that the people who wrote these accounts were mostly simple individuals who had no literary pretensions, and the works themselves were things like diaries and records intended to be passed down in families. One item that he says was written to *"create the desired impression, possibly at the expense of strict representational accuracy"* is the following:

> Due to war, pestilence, price rise and famine, our people are reduced to such an extent, that it will be difficult for our descendants to believe it.

---

[16] Andrew P. Williams, 'Shifting Signs: Increase Mather and the Comets of 1680 and 1680': http://extra.shu.ac.uk/emls/01-3/willmath.html

[17] http://www.weatherfriend.com/astronomy/comet/cometlist.html

Now, one has to keep in mind the meaning of the word 'pestilence' as we discussed already in a previous chapter. Jon Arrizabalaga, in his article included in *Practical Medicine from Salerno to the Black Death*, discusses the etiology of this word and how it was understood by the peoples of the time. He writes:

> The emphasis placed on celestial causes of the 'pestilence' by the different physicians studied here varied quite widely. ... In 1340 Augustine of Trent, a friar eremite of St Augustine, justified having written a medical and astrological work on a 'pestilence of diseases' happening everywhere in Italy, because of physicians' ignorance about the roots of diseases; this fact was considered by him 'a pestiferous mistake involving many physicians', and he blamed it on their *'ignorance of astronomy'*. ...
>
> Works from other geographical areas assigned a more relevant role to celestial causes in the genesis of the 'pestilence.' ...
>
> Jacme d'Agramaont ... said nothing concerning the term *epidímia*, but he extensively developed what he meant by *pestilència*. He gave this latter term a very peculiar etymology, in accordance with a form of knowledge established by Isidore of Seville (570–636) in his *Etymologiae*, which came to be widely accepted throughout Europe during the Middle Ages. He split the term *pestilència* up into three syllables, each having a particular meaning: *pes* (= *tempesta*: 'storm', 'tempest'), *te* (= *temps*: 'time'), and *lència* (= *clardat*: 'brightness', 'light'); hence, he concluded, the *pestilència* was *'the time of tempest caused by light from the stars.'* [emphasis added] (Garcia-Ballester et al. 1994, 252, 253, 244)

And so, we have a better idea of what our German diarist meant when he said: *"Due to war, pestilence, price rise and famine, our people are reduced to such an extent, that it will be difficult for our descendants to believe it."*

On page 5 (101) of Mortimer's paper, we read that a young officer at the time of the sack of Magdeburg in 1631 wrote in his memoirs:

> [A] grand storm-wind picked up, the town was inflamed at all possible places, so that even little aid (rescue) was of help (appreciated). ... then I saw the whole town of Magdeburg, except dome, monastery and New Market, lying in embers and ashes, which raged only about 3 or 3 ½ hours, from which I deduced God's strange omnipotence and punishment.

A 'grand storm-wind' and a town that was 'inflamed' all over at once, and burned to cinders in 3.5 hours? Perhaps the reader will like to go back and re-read the description of how an overhead cometary explosion would manifest, quoted at the end of chapter three.

Note the date of the above event: 1631. As it happens, there were other mysterious things happening on the planet at that time. In her book *Comets and Popular Culture and the Birth of Modern Cosmology*, Sara J. Schechner writes:

> Comets, like other marvels, were exploited by polemicists in prodigy books. In 1661–1662, for example, radical English dissenters ... published sensationalist reports of prodigies, including comets, which gloomily greeted the restoration of Charles II. ... There were no fewer than twenty-five apparitions visible in seventeenth century Europe, and these comets made frequent appearances in the polemical broadsheets and chapbooks hawked in the marketplaces ... (Schechner 1999, 69, 72)

Comets were apparently a common occurrence during this time. One of these tracts shows comets in 1680, 1682, and 1683. Another shows five comets between 1664 and 1682. Another talks about comets of 1618. A tract entitled 'The Signs of The Times' shows a bunch of prodigies that accompanied comets. Schechner writes:

> All these outbursts were concerned with specific political quarrels. Some pamphleteers, however, raised themselves above the local rough water to examine a larger vista. They thought they saw a fast-approaching end to the world and their works adopted an apocalyptic tone. The comet of 1580 confirmed Francis Shakelton in his opinion that the Day of Judgment was near at hand. ... Although Regiomontanus and others agreed that *1588 would be a year of great revolutions and world mutations*, Jesus had yet to reappear when William Lilly viewed the comets of 1664, 1665, and 1673 as tokens of the beginning of the end. In comets like that of 1680, Ezerel Tonge, Christopher Ness, and others saw the great "northern star," the messianic herald of the last days predicted by the sibyl Tiburtina and Tycho Brahe.
>
> Panic and joy were heightened by the great conjunction of Jupiter and Saturn in the fiery trigon in 1682, which came on the heels of a comet's apparition. While great conjunctions take place every twenty years, this one was part of an astrologically profound series of conjunctions that commenced with the climacteric (or "maximum") conjunction at the close of the sixteenth century. By definition, climacteric conjunctions occurred only every eight hundred years when the great conjunction of Jupiter and Saturn returned to the sign of Aries and to the fiery trigon ... It was widely reported by the popular press that Tycho Brahe, Johannes Kepler, and Johann Heinrich Alsted ... correlated historical periods with climacteric conjunctions and believed that they portended great mutations and reformations of the spiritual and secular regimes. Tycho Brahe reckoned that all odd-numbered maximum conjunctions were auspicious and urged people to look forward to the period of the sabbatical or seventh climacteric

conjunction since the world's Creation, which he believed would follow the conjunction in Aries in 1583. During the conjunctions in Leo in October 1682, the planets allegedly would be in the same configuration as they had been at the beginning of the world. Alsted believed this might be the last conjunction of the present world and *publicly announced that the Millennium would commence in 1694.*

By itself, the great conjunction in the fiery trigon was a serious matter, but its power was corroborated by several other signs. Mars joined Jupiter and Saturn in 1682. There was a solar eclipse. But most critically, the great conjunction was ushered in by the comets of 1680 and 1682, and the former was said to have been unrivaled in eight hundred years. Many thought the comets augured the Apocalypse ... the end of the world ... In sensationalist street literature, radical pamphleteers took advantage of these comets. ...

... At the restoration, *the Crown cracked down on the almanacs of Lilly and others, blaming them for fomenting insurrection and irreligion during the Civil War and Interregnum.* [emphasis added] (Schechner 1999, 80, 82)

The author next discusses the major controls put in place at this point to stamp out the popular discussion of predictions, interpretations of 'signs in the skies'. So we can understand how so much of this period of 'panic' when 'governments fell' was covered up. Based on the number of pamphlets and broadsides, it must have been a really crazy time and everybody was thinking the world was going to end. *But*, as we go through this description, we find a most interesting item that relates to what our young officer witnessed at the fall of Magdeburg:

The sunny disposition of the weather during the coronation [of Charles II] was seen as the fulfillment of a prophecy. In 1630, at the time of Charles' birth, a noonday star or rival sun allegedly had appeared in the sky. Aurelian Cook in *Titus Britannicus* explained its import:

'As soon as Born, Heaven took notice of him, and ey'd him with a Star, appearing in defiance of the Sun at Noon-day ...'

For Cook, the extra sun announced that Charles ruled by divine right. Moreover, the timing of Charles' entry into London on his birthday was politically calculated to fulfill what had been portended at his birth. The point was grasped immediately by Abraham Cowley (1518–1667), poet, diplomat, and spy for the court during its exile:

'No *Star* amongst ye all did, I beleeve,
Such vigorous assistance give,
As that which thirty years ago,

> At *Charls* his *Birth*, did, in despight
> Of the proud *Sun's* Meridian Light,
> His future *Glories*, this *Year* foreshow.'

Edward Matthew devoted an entire book to the fulfillment of the prophecy, declaring Charles "ordained to be the most *Mighty Monarch* in the *Universe*".

Charles' return was seen as a rebirth for England and duly recorded by a special act in the statute book, which proclaimed that 29 May was "the most memorable Birth day not only of his Majesty both as a man and Prince, but likewise as an actual King …" (Schechner 1999, 84)

So, a 'second sun' was seen on and around 29 May 1630, and on 30 May 1631, one year later, Magdeburg fell as described by our young officer.

The standard historical description of the Fall of Magdeburg goes pretty much as follows:

> The fall of Magdeburg horrified Europe. The city had been starved and then was bombarded unmercifully. The artillery shelling grew so bad, the town caught on fire. Over 20,000 of the citizens perished in the siege and the cataclysm that ended it. The city itself was burned to the ground. The cruel and pointless devastation marked a new low, an act abhorred by a generation well accustomed to horrors.[18]

The war was to continue for 17 more years. Twenty or 30 years later a lot of new comets showed up, and I used to think that this 'second sun' seen at the time of the birth of Charles II may have been an appearance of our sun's twin in the far reaches of the solar system. However, with the scientific information provided by Clube and Napier *et al*, I have changed my view.

In any event, we begin to see why Clube wrote:

> [W]hen the prospect of these global catastrophes recurs, such is the nerve-racking tension aroused in mankind that the principal leaders of civilization have long been in the habit of dissembling as to their cause (and likelihood) simply in order to preserve public calm and avoid the total breakdown of civil affairs. …
>
> The Christian, Islamic and Judaic cultures have all moved since the European Renaissance to adopt an unreasoning anti-apocalyptic stance, apparently unaware of the burgeoning science of catastrophes. History, it now seems, is repeating itself: it has taken the Space Age to revive the Platonist voice of reason but it emerges this time within a modern anti-fundamentalist, anti-apocalyptic tradi-

---

[18] http://www.boisestate.edu/courses/reformation/germany/30yw.shtml

tion over which governments may, as before, be unable to exercise control. The logical response is perhaps a commitment on the part of government to the voice of reason and a decision to eliminate all signs as well as perpetrators of cosmic catastrophes in order to appease a public not too far given to rabid uniformitarianism. *Cynics ... would say that we do not need the celestial threat to disguise Cold War intentions; rather we need the Cold War to disguise celestial intentions!* [emphasis added]

We see that the events of those times have been covered up and/or forgotten, for the most part in their historical context.

Long after the event, John Dryden suggested that the comets of 1664 and 1665 were related to the sun that was seen at the birth of Charles II. He described this apparition as *"That bright companion of the sun ..."*

After the Thirty Years War was over, comets were associated with witches and both were written off as superstition by the Protestants who prided themselves on having ushered in the scientific age. Andrew C. Fix, professor of History at LaFayette College, PA, writes:

Blathasar Bekker was a minister in the Dutch Reformed church first in Friesland and then in Holland. He was educated in philosophy and theology at the northern Dutch universities of Groningen and Franeker, becoming a Doctor of Theology at Franeker. Influenced by Cartesian philosophy, he was an important critic of belief in witchcraft in his book *De Betoverde Weerld* (*The World Bewitched*) in which he argued against the possibility that disembodied spirits could contact, influence, or do evil to human beings, and thus against the possibility of witchcraft. ...

After writing a work critical of the terrestrial influence of comets Bekker became interested in other popular superstitions including witchcraft and sorcery. He approached these topics from the point of view of a Reformed minister upholding the power and earthly influence of God against the supposed power of witches and spirits. ...

In the discussions around the Sabbath, the earthly effects of comets, and witchcraft Bekker was motivated in part by Cartesian rationalism, in part by his Calvinist idea of God's omnipotence, and in part by his view of Scriptural exegesis, which included the doctrine of accommodation, the idea that God had in some places accommodated his holy language to the limited understandings of men. In volume one of *The World Bewitched* Bekker maintained that belief in the Devil and evil spirits as well as in such things as fortune telling, sorcery, and witchcraft were originally pagan beliefs founded upon ignorance, prejudice, and fear that had over time crept into the Catholic church and even into Bekker's own Reformed tradition. In volume two of the work Bekker applied Cartesian dualism to argue that the material and spiritual worlds could not interact with each other outside man and therefore spirits without bodies such as the Devil could have no influence or effect on people.[19]

And so it was that records of the phenomena of that time as having any impact on earthly matters have been explained away, covered up, dismissed, consigned to superstition and 'cults'.

---

[19] Andrew C. Fix, 'Angels, Devils, and Evil Spirits in Seventeenth-Century Thought: Balthasar Bekker and the Collegiants': http://www.historicum.net/themen/hexenforschung/lexikon/alphabethisch/a-g/art/Bekker_Balthas/html/artikel/5576/ca/aaff064c57/

6. Comet Biela
and Mrs. O'Leary's Cow

The next period of cometary activity that Clube refers to is that which encompassed the American Revolution (1775–1783) and the French Revolution (1789–1799) and the mid-nineteenth century crisis. I'm going to skip the two revolutions for the moment and go directly to the mid-nineteenth century period because it is so intensely interesting and leads us into the main topic of this chapter.

In trying to find some details about the mid-nineteenth century crisis mentioned above, a whole lot of details turn up that I'm sure we all learned in history class in school, but they are never put together in such a way as to make them look as interesting as they actually were. What happened then was, of course, the 'Industrial Revolution'. But, just like the Renaissance, this overlapped several other interesting events.

The Industrial Revolution and the rise of capitalism began, more or less, toward the end of the eighteenth century. The nineteenth century was a turbulent epoch beginning with a stock market crash in 1825 then moving on to the Panic of 1847, a collapse of British financial markets associated with the end of the 1840s railroad boom. The crisis of 1847 could have been more disastrous except that it was cut short by economic revival following the California gold strike of 1849.

After a period of prosperity, there began a series of wars and revolutions. There was the first Italian War for Independence in 1857, and then the American Civil War of 1861, the Polish Insurrection of 1863, Napoleon the Second's Mexican adventure and the campaign against Denmark in 1864, which started the Prussian Wars led by Bismarck. Bismarck attacked Austria in 1866 and won a victory over France in 1871. Then, there was the Republican uprising in Spain which toppled Queen Isabella from the throne. Finally, there was the last of Louis Napoleon's adventures, which culminated in the crashing of the Empire in 1871.

There was civil war in France following the downfall of the Second Napoleon, and the people (Paris Communards) seized power. They were soon crushed, order was restored in the Third Republic, and the revolutionary tide receded for the rest of the century.

It is interesting to consider the other events that were occurring at this time. Industrial capitalism was being spread with missionary zeal everywhere. Western investors roamed the globe looking for openings to establish trade and to invest in anything that could be bought or sold. In the process, millions of people were redistributed in the greatest mass migrations in history from the Old World to the New. Science became the handmaiden of industry and capitalism. The volume of world trade was 1.75 billion dollars in 1830 and it rose to 3.6 billion in 1850, skyrocketing to 9.4 billion in 1870.

So, Clube is right. For about twenty-five years, the entire Western world was bubbling cauldron of war and revolution and people taking advantage of wars and revolution to make money. When it was all over, the imperial powers of Europe that were to rule the world until 1914 were firmly ensconced. More than that, the United States as a federal, capitalist entity, had been forged at Appamattox.

There were obviously other things going on at that time. In the period from 1830 to 1860 there was apparently an enormous upsurge in religious fervor. The imminent return of Christ was being predicted everywhere. Manuel de Lacunza, a Catholic priest in South America wrote (under the pen name of Juan Josafa Ben-Ezra) a book entitled *The Coming of Messiah in Glory and Majesty*, which was published in Spain in 1812. He believed that Jesus was coming very, very soon. William Miller (Seventh-Day Adventists) declared that Christ was coming and predicted 1844 as the date. Edward Irving of England and Johann Bengel in Germany almost simultaneously came to the conclusion that the prophecies of Daniel pointed to the time of the end being right then; Mason in Scotland, Leonard H. Kelber in Germany and many, many others preached about the Second coming. Spiritualist Andrew Jackson Davis gave 157 lectures in 1845 about the new era, which Edgar Allen Poe attended regularly. The spiritualism craze began with the Fox sisters in 1848. Mourant Brock, of the Church of England, noted that the craze for eschatology had spread through all of Europe and extended to India. (See *The Story of Prophecy* by Henry James Forman.)

As Clube notes, this religious fervor parallels cosmic events.

In 1843, there appeared one of the greatest comets of history. The Great Comet of 1843, formally designated C/1843 D1 and 1843 I, was discovered on 5 February 1843 and rapidly brightened. It was a member of the Kreutz Sungrazers, a family of comets resulting from the breakup of a parent comet (X/1106 C1) into multiple fragments in about 1106. These comets pass extremely close to the Sun – within a few solar radii – and this is why they often become very bright.

C/1843 D1 moved rapidly toward an incredibly close perihelion of less than 830,000 km on 27 February 1843, at which time it could be seen in broad daylight just a degree away from the sun. It swung around and passed close to earth on 6 March 1843, and seemed to manifest its greatest brilliance the following day. It was last observed on 19 April 1843. At that time, this comet had passed closer to the sun than any other known object. *The American Journal of Science* and *The New York Tribune* devoted special sections to this comet at the time. You could say that 'comet fever' was pandemic.

The Great Comet of 1843 – still unnamed – developed a tail over two astronomical units in length, the longest known cometary tail until measurements in 1996 showed that Comet Hyakutake's tail was almost twice as long.

In 1857, an anonymous German astrologer predicted that a comet would strike the earth on June 13 of that year. The impending catastrophe became the talk of all of Europe. The French astronomer, Jacques Babinet, tried to reassure people by stating that a collision between the earth and a comet would do no harm. He compared the impact to *"a railway train being hit by a fly"*. His words, apparently, had little effect. The Paris correspondent for the American journal, *Harper's Weekly*, wrote:

> Women have miscarried; crops have been neglected; wills have been made; comet-proof suits of clothing have been invented; a cometary life insurance company (premiums payable in advance) has been created ... all because an almanac maker ... thought proper to insert, under the week commencing June 13, 'About this time, expect a comet'.

Let's back up just a minute here, to 1826. In 1826, comet 3D/Biela was discovered by Wilhelm von Biela. It has become known as Comet Biela or Biela's Comet. This comet had been first seen in 1772 by Charles Messier and again in 1805 by Jean-Louis Pons. It was von Biela who discovered it in its 1826 perihelion approach (on 27 February) and calculated its orbit, discovering it to have a period of 6.6 years which is why it was named after him and not Messier or Pons. It was only the third comet (at the time) found to be periodic, after the famous comets Halley and Encke. French astronomer M. Damoiseau subsequently calculated its path and announced that on its next return the comet would cross the orbit of the earth, within twenty thousand miles of its track, and about one month before the earth would arrive at the same spot.

When the comet came in 1832, the earth did, indeed, miss it by one month. It returned again in 1839 and 1846. In its 1846 appearance, the comet was observed to have broken up into two pieces. It was observed again in 1852 with the two parts being 1.5 million miles apart. Each fragment had a head and tail of its own. The comet did not come in 1845, 1859, or 1866. *The Edinburgh Review* notes about this strange state of affairs:

The puzzled astronomers were left in a state of tantalizing uncertainty as to what had become of it. At the beginning of the year 1866 this feeling of bewilderment gained expression in the Annual Report of the Council of the Royal Astronomical Society. The matter continued, nevertheless, in the same state of provoking uncertainty for another six years. The third period of the perihelion passage had then passed, and nothing had been seen of the missing luminary. But on the night of November 27, 1872, night-watchers were startled by a sudden and a very magnificent display of falling stars or meteors, of which there had been no previous forecast ... (quoted in Donnelly 1883, 410)

The meteors were radiating from the part of the sky where the comet had been expected to cross in September. In other words, the trajectory was the same, and the earth intersected it, but the velocity was somewhat altered. *The American Journal of Science* said they fell like snowflakes. Professor Olmstead, a mathematician at Yale University, estimated 34,640 shooting stars per hour. *The New York Journal of Commerce* wrote that no philosopher or scholar has ever recorded an event like this. These meteors became known as the Andromedids or 'Bielids' and it seems apparent that they indicated the death of the comet. The meteors were seen again on subsequent occasions for the rest of the 19th century, but have now faded away.

Is that all there is to that? Maybe not. The following passages of accounts are extracted from Ignatius Donnelly's *Ragnarok: The Age of Fire and Gravel* (italics in original):

In the year 1871, on Sunday, the 8th of October, at half past nine o'clock in the evening, events occurred which caused the death of hundreds of human beings, and the destruction of millions of property, and which involved three different States of the Union in the wildest alarm and terror.

The summer of 1871 had been excessively dry; the moisture seemed to be evaporated out of the air; and on the Sunday above named the atmospheric conditions all through the Northwest were of the most peculiar character. The writer was living at the time in Minnesota, hundreds of miles from the scene of the disasters, and he can never forget the condition of things. There was a parched, combustible, inflammable, furnace-like feeling in the air, that was really alarming. It felt as if there were needed but a match, a spark, to cause a world-wide explosion. It was weird and unnatural. I have never seen nor felt anything like it before or since. Those who experienced it will bear me out in these statements.

At that hour, half past nine o'clock in the evening, *at apparently the same moment*, at points hundreds of miles apart, in three different States, Wisconsin, Michigan, and Illinois, fires of the most peculiar and devastating kind broke out, so far as we know, by spontaneous combustion.

In Wisconsin, on its eastern borders, in a heavily timbered country, near Lake Michigan, a region embracing *four hundred square miles*, extending north from Brown County, and containing Peshtigo, Manistee, Holland, and numerous villages on the shores of Green Bay, was swept bare by an absolute whirlwind of flame. There were *seven hundred and fifty people killed outright*, besides great numbers of the wounded, maimed, and burned, who died afterward. More than three million dollars' worth of property was destroyed. [See "History of the Great Conflagration," Sheahan & Upton, Chicago, 1871, pp. 393, 394, etc.]

It was no ordinary fire. I quote:

"At sundown there was a lull in the wind and comparative stillness. For two hours there were no signs of danger; but at a few minutes after nine o'clock, and by a singular coincidence, *precisely the time at which the Chicago fire commenced*, the people of the village heard a terrible roar. It was that of a tornado, crushing through the forests. *Instantly the heavens were illuminated with a terrible glare. The sky*, which had been so dark a moment before, *burst into clouds of flame*. A spectator of the terrible scene says the fire did not come upon them gradually from burning trees and other objects to the windward, but the first notice they had of it was *a whirlwind of flame in great clouds from above the tops of the trees*, which fell upon and entirely enveloped everything. The poor people inhaled it, or the intensely hot air, and fell down dead. This is verified by the appearance of many of the corpses. They were found dead in the roads and open spaces, *where there were no visible marks of fire near by, with not a trace of burning upon their bodies or clothing*. At the Sugar Bush, which is an extended clearing, in some places four miles in width, corpses were found in the open road, between fences only slightly burned. *No mark of fire was upon them; they lay there as if asleep*. This phenomenon seems to explain the fact that so many were killed in compact masses. They seemed to have huddled together, in what were evidently regarded at the moment the safest places, *far away from buildings, trees, or other inflammable* material, and there to have died together." [p. 372]

Another spectator says:

"Much has been said of the intense heat of the fires which destroyed Peshtigo, Menekaune, Williamsonville, etc., but all that has been said can give the stranger but a faint conception of the reality. The heat has been compared to that engendered by a flame concentrated on an object by a blow-pipe; but even that would not account for some of the phenomena. For instance, we have in our possession a copper cent taken from the pocket of a dead man in the Peshtigo Sugar Bush, which will illustrate our point. *This cent has been partially fused*, but still retains its round form, and the inscription upon it is legible. Others, in the same pocket, were partially melted, and yet *the clothing and the body of the man were not even singed*. We do not know in what way to account for this, unless, as is asserted by some, the tornado and fire were accompanied by electrical phenomena." [p. 373]

"It is the universal testimony that the prevailing idea among the people was, that the last day had come. Accustomed as they were to fire, nothing like this had ever been known. They could give no other interpretation to this ominous roar, this *bursting of the sky with flame, and this dropping down of fire out of the very heavens*, consuming instantly everything it touched.

"No two give a like description of the great tornado as it smote and devoured the village. It seemed as if 'the fiery fiends of hell had been loosened,' says one. 'It came in great sheeted *flames from heaven*,' says another. 'There was a pitiless rain of fire and SAND.' 'The atmosphere was all afire.' Some speak of *'great balls of fire unrolling and shooting forth in streams.'* The fire leaped over roofs and trees, and ignited whole streets at once. No one could stand before the blast. It was a race with death, above, behind, and before them." [p. 374]

A civil engineer, doing business in Peshtigo, says:

"The heat increased so rapidly, as things got well afire, that, *when about four hundred feet from the bridge and the nearest building*, I was obliged to lie down behind a log that was aground in about two feet of water, and by going under water now and then, and holding my head close to the water behind the log, I managed to breathe. There were a dozen others behind the same log. If I had succeeded in crossing the river and gone among the buildings on the other side, probably I should have been lost, as many were."

…

In Michigan, one Allison Weaver, near Port Huron, determined to remain, to protect, if possible, some mill-property of which he had charge. He knew the fire was coming, and dug himself a shallow well or pit, made a thick plank cover to place over it, and thus prepared to bide the conflagration.

I quote:

"He filled it nearly full of water, and took care to saturate the ground around it for a distance of several rods. Going to the mill, he dragged out a four-inch plank, sawed it in two, and saw that the parts tightly covered the mouth of the little well. 'I calculated it would be touch and go,' said he, 'but it was the best I could do.' At midnight he had everything arranged, and the roaring then was awful to hear. The clearing was ten to twelve acres in extent, and Weaver says that, for two hours before the fire reached him, there was a constant flight across the ground of small animals. As he rested a moment from giving the house another wetting down, a horse dashed into the opening at full speed and made for the house. Weaver could see him tremble and shake with excitement and terror, and felt a pity for him. After a moment, the animal gave utterance to a snort of dismay, ran two or three times around the house, and then shot off into the woods like a rocket. …

"Not long after this the fire came. Weaver stood by his well, ready for the emergency, yet curious to see the breaking-in of the flames. The roaring increased in volume, the air became oppressive, a cloud of dust and cinders came showering down, and he could see the flame through the trees. It did not run along the ground, or leap from tree to tree, but it came on like a tornado, *a sheet of flame reaching from the earth to the tops of the trees*. As it struck the clearing he jumped into his well, and closed over the planks. He could no longer see, but he could hear. He says that the flames made no halt whatever, or ceased their roaring for an instant, but he hardly got the opening closed before the house and mill were burning tinder, and both were down in five minutes. The smoke came down upon him powerfully, and his den was so hot he could hardly breathe.

"He knew that the planks above him were on fire, but, remembering their thickness, he waited till the roaring of the flames had died away, and then with his head and hands turned them over and put our fire by dashing up water with his hands. Although it was a cold night, and the water had at first chilled him, the heat gradually warmed him up until he felt quite comfortable. He remained in his den until daylight, frequently turning over the planks and putting out the fire, and then the worst had passed. The earth around was on fire in spots, house and mill were gone, leaves, brush, and logs were swept clean away as if shaved off and swept with a broom, and nothing but soot and ashes were to be seen." [p. 390]

In Wisconsin, at Williamson's Mills, there was a large but shallow well on the premises belonging to a Mr. Boorman. The people, when cut off by the flames and wild with terror, and thinking they would find safety in the water, leaped into this well. "The relentless fury of the flames drove them pell-mell into the pit, to struggle with each other and die – some by drowning, and others by fire and suffocation. None escaped. *Thirty-two bodies were found there.* They were in every imaginable position; but the contortions of their limbs and the agonizing expressions of their faces told the awful tale." [p. 386]

James B. Clark, of Detroit, who was at Uniontown, Wisconsin, writes:

"The fire suddenly made a rush, like the flash of a train of gunpowder, and swept in the shape of a crescent around the settlement. It is almost impossible to conceive *the frightful rapidity of the advance of the flames*. The rushing fire seemed to eat up and annihilate the trees."

They saw a black mass coming toward them from the wall of flame:

"It was a stampede of cattle and horses thundering toward us, bellowing, moaning, and neighing as they galloped on; rushing with fearful speed, their eyeballs dilated and glaring with terror, and every motion betokening delirium of fright. Some had been badly burned, and must have plunged through a long space of flame in the desperate effort to escape. Following considerably behind came a solitary horse, panting and snorting and nearly exhausted. He was saddled and bridled, and, as we first thought, had a bag lashed to his back. As he came up we were startled at the sight of a young lad lying fallen over the animal's neck, the bridle wound around his hands, and the mane being clinched by the fingers. Little effort was needed to stop the jaded horse, and at once release the helpless boy. He was taken into the house, and all that we could do was done; but he had inhaled the smoke, and was seemingly dying. Some time elapsed and he revived enough to speak. He told his name–Patrick Byrnes–and said: 'Father and mother and the children got into the wagon. I don't know what became of them. Everything is burned up. I am dying. Oh! Is hell any worse than this?'" [p. 383]

... When we leave Wisconsin and pass about two hundred and fifty miles eastward, over Lake Michigan and across the whole width of the State of Michigan, we find much the same condition of things, but not so terrible in the loss of life. Fully *fifteen thousand people were rendered homeless by the fires*; and their food, clothing, crops, horses, and cattle were destroyed. Of these five to six thousand were burned out *the same night that the fires broke out in Chicago and Wisconsin.* The total destruction of property exceeded one million dollars; not only villages and cities, but whole townships, were swept bare.

But it is to Chicago we must turn for the most extraordinary results of this atmospheric disturbance. It is needless to tell the story in detail. The world knows it by heart. ... I have only space to refer to one or two points.

The fire was spontaneous. The story of Mrs. O'Leary's cow having started the conflagration by kicking over a lantern was proved to be false. It was the access of gas from the tail of Biela's comet that burned up Chicago!

The fire-marshal testified:

"I felt it in my bones that we were going to have a burn."

He says, speaking of O'Leary's barn:

"We got the fire under control, and it would not have gone farther; but the next thing I knew they came and told me that St. Paul's church, *about two squares north, was on fire*." [p. 163]

They checked the church-fire, but –

"The next thing I knew the fire was in Bateham's planing-mill."

A writer in the New York "Evening Post" says he saw in Chicago "buildings far beyond the line of fire, and *in no contact with it, burst into flames from the interior*."

It must not be forgotten that the fall of 1871 was marked by extraordinary conflagrations in regions widely separated. On the 8th of October, *the same day* the Wisconsin, Michigan, and Chicago fires broke out, the States of Iowa, Minnesota, Indiana, and Illinois were severely devastated by prairie-fires; while terrible fires raged on the Alleghenies, the Sierras of the Pacific coast, and the Rocky Mountains, and in the region of the Red River of the North.

"The Annual Record of Science and Industry" for 1876, page 84, says: "For weeks before and after the great fire in Chicago in 1872, great areas of forest and prairie-land, both in the United States and the British Provinces, were on fire."

The flames that consumed a great part of Chicago were of an unusual character and produced extraordinary effects. They absolutely melted the hardest building-stone, which had previously been considered fire-proof. Iron, glass, granite, were fused and run together into grotesque conglomerates, as if they had been put through a blast-furnace. No kind of material could stand its breath for a moment.

I quote again from Sheahan & Upton's work:

"The huge stone and brick structures melted before the fierceness of the flames as a snow-flake melts and disappears in water, and almost as quickly. Six-story buildings would take fire and *disappear for ever from sight in five minutes by the watch*. ... The fire also doubled on its track at the great Union Depot and burned half a mile southward in the very teeth of the gale – a gale which blew a perfect tornado, and in which no vessel could have lived on the lake. ... *Strange, fantastic fires of blue, red, and green played along the cornices of buildings*." [pp. 85, 86]

Hon. William B. Ogden wrote at the time:

"The fire was accompanied by the fiercest tornado of wind ever known to blow here." [p. 87]

"The most striking peculiarity of the fire was its intense heat. Nothing exposed to it escaped. Amid the hundreds of acres left bare there is not to be found a piece of wood of any description, and, *unlike most fires, it left nothing half burned*. ... The fire swept the streets of all the ordinary dust and rubbish, consuming it instantly." [p. 119]

The Athens marble burned like coal!

A depiction of the great fire of Chicago, 9 October 1871.

"The intensity of the heat may be judged, and the thorough combustion of everything wooden may be understood, when we state that in the yard of one of the large agricultural-implement factories was stacked some hundreds of tons of pig-iron. This iron was two hundred feet from any building. To the south of it was the river, one hundred and fifty feet wide. No large building but the factory was in the immediate vicinity of the fire. Yet, so great was the heat, that *this pile of iron melted and run, and is now in one large and nearly solid mass.*" [p. 121]

The amount of property destroyed was estimated by Mayor Medill at one hundred and fifty million dollars; and the number of people rendered houseless, at one hundred and twenty-five thousand. Several hundred lives were lost. (Donnelly 1883, 413–423)

No doubt this story came to the attention of Victor Clube.

Ten years later, there was the Great Comet of 1881 (C/1881 K1), discovered by the Australian amateur astronomer, John Tebbutt. All we hear about this comet nowadays is that it was one of the first comets photographed and studied scientifically. However, this comet following so closely on the events of the previous ten years obviously got a few people thinking.

Minnesota Congressmen Ignatius Donnelly, who had already stated that *he thought the Great Chicago Fire had been caused by cometary debris*, published *Ragnarok*, wherein he proposed that a giant comet had passed close to the earth in past ages.

The intense heat from the comet had set off huge fires that raged across the face of the globe. He suggested that the comet had dumped vast amounts of dust on the earth, triggered earthquakes, leveled mountains, and initiated the ice age. He even explained some of the miracles of the Bible in terms of his comet, proposing that the standing-still of the sun at the command of Joshua was possibly a tale commemorating this event. Donnelly's readers were thrilled by his descriptions of the 'glaring and burning monster' in the sky, scorching the planet with unearthly heat and shaking the land with 'thunders beyond all thunders'.

Possibly inspired by Donnelly (not to mention what was obviously going on in the heavens), Camille Flammarion wrote *The End of the World* in 1893 in which he recounted a fictional collision between the earth and a comet fifty times its size. Flammarion's lurid prose ensured that his book was an immediate sensation. (Flammarion, it should be noted, was a friend and associate of, and greatly influenced by, Allan Kardec, the French pedagogue, medical student, linguist and researcher of 'spirit communications'. He was also a friend of Jules Violle, the probable true identity of the legendary alchemist, Fulcanelli.[20])

---

[20] See: *Fulcanelli – His True Identity Revealed* by Patrick Rivière.

With this fascinating diversion into history in mind, I think that Victor Clube's writing will make a whole lot more sense. Returning to his narrative:

> The fact of a perceived danger at these epochs, signified historically by a global rise in eschatological concern, is now understood in various academic quarters as marking some kind of physical dislocation (climate? disease?) which causes economic and social activity to be widely deranged, even to the point of collapse of civilized society, leading then to revolution, mass migration and war, amplified on a global scale.

> The occasions of such breakdowns in civilization are of course a matter of serious concern and their systematic study has been taken up in America (and elsewhere) at such institutes as the Center for Comparative Research in History, Society and Culture at the University of California, Davis (Goldstone, 1991). To the "enlightened" however, the eschatology remains an anomaly and secure connections with celestial inputs have generally still to be made. We should recall however that many, as usual on these occasions of breakdown, would see "blazing stars threatening the world with famine, plague and war; to princes' death; to kingdoms many curses; [and] to all estates many losses ..."

> The three earliest of these epochs are of course the periods of Inquisition and of the great European witch-hunts (which spilled over to America) when ecclesiastical and secular administrators alike would discourage any (astrological) notion that the celestial sphere interfered with terrestrial affairs. The separate stories of scientific revolutionaries like Copernicus, Kepler, Bruno, Galileo and Newton now bear witness to the ferocity with which the most acceptable cosmic viewpoint (of the time) was imposed. Indeed, these separate stories are still being adjusted and Newton, it is now realised, was constrained by his times to work under conditions of rather considerable censorship.

> The acceptable part of his scientific output was of course published and has proved its worth repeatedly over 300 years. The unacceptable part however dealt with "blazing stars" and eschatology and remained unpublished for some 250 years. One of the first to examine this material (Keynes 1947) was so taken aback by the contrast as to dub Newton not so much "the first of the age of reason" as "the last of the magicians, the last of the Babylonians and Sumerias". Thus it was the Founding Fathers of the Royal Society in Restoration England who hit upon the "enlightened" step of deriding the cosmic threat and public anxiety; and it is not without significance today that English-speaking nations ultimately stood firm and prospered as others faltered at the last and briefest of the above epochs (Goldstone, *loc cit*).

> Accordingly, it is largely an Anglo-Saxon "achievement" that cosmic catastrophes were absolutely discarded and the scientific principle of uniformitarianism was put in place between 200 and 150 years ago.

If short-period bombardment of our planet by comets or comet dust is a reality (as it increasingly appears to be); and the effects of such an event are deleterious in the extreme; and if we are in fact overdue for a repeat performance of such a visitation (which also appears to be the case); what effect might public awareness of this have on the status quo on the planet at present? Would the bogus 'War on Terror' not become instantly obsolete and would people across the planet not immediately demand that their political leaders reassess priorities and take whatever action possible to mitigate the threat? And if those political leaders refused to do so and it became known that that this grave threat to the lives of billions was long-standing and common knowledge among the political elite (with all that that implies), what then? Revolution? One last hurrah before the sixth extinction?

Who knows? We only know that this knowledge, in its fullest explication, is being suppressed and marginalized. The reasons for the psychological games and ploys may be interesting to investigate, so that is what we will look at shortly: why is humanity so deaf, dumb and blind?

7. Tunguska, the Horns of the Moon and Evolution

During the course of writing this book, my household watched *Super Comet: After the Impact*, a Discovery Channel special that takes the comet that wiped out the dinosaurs and places it in a modern setting. The fictionalized drama (admittedly quite cheesy) follows the struggles of several individuals or groups, before, during, and after the impact, to show how people would react to such a global cataclysm. The filmmakers use the same type of cometary body assumed to have caused the extinction of the dinosaurs – the same size, same impact location – and utilize all the computer modeling they have done on this past event to try to show what might happen (and what they think happened then). The end result is not terribly creative and suggests that they really don't know all the effects of such an impact. Rather, they are just patching things together from what little is known about that one impact, some (or much) of which may be just speculation, though I'm sure that there is some good science used as a basis for the production.

The cheesiest part of this 'docu-drama' was, of course, the depicted foibles of the humans experiencing the event. But, in a way, even those depictions were useful. One character who simply couldn't grasp the nature of the event, kept traveling 'home' (which happened to be the site of the impact) even when it was clear that there was no home left. His emotions basically drove him to his own death. Other people continued to act as if the world was still the same place and suffered thereby, though they learned to cope. What was clearly evident was that it was lack of knowledge about such events that was the chief problem for all of them.

During the course of the show, one of the experts made the remark 'when it happens', as though he – and the rest of them – knew for a fact that this was on the agenda for our near future. The very fact that so many scientists are working on these problems, including a large number of them studying the possible human reactions and behaviors and how to deal with masses of people, should warn us that there is something they aren't telling the masses in the headlines of our daily newspapers, though certainly they are 'testing' public reactions with shows such as *Super Comet: After the Impact*.

This show highlights what we have already noted: the difference between the American school of asteroid impacts that happen only at millions-of-years intervals and the British school, which posits that showers of much smaller objects occur with great frequency in between those millions-of-years events. Now, coming at this from a different angle, studies of the history of the earth via various scientific methods show us that there are relatively long periods of 'evolution' punctuated by rapid, overwhelming changes we call catastrophes. Many scientists have noted the periodicity of these punctuational events. What no one seems to know for sure is the mechanism that induces these definitely periodic catastrophes.

It is suggested that the periodicity of these events relates to galactic cycles and there is good evidence for this view presented by Victor Clube and Bill Napier in their book *The Cosmic Winter*. They suggest that galactic tides induct giant comets into our solar system and it is their disintegration products that interact strongly and directly with the earth with variable results at different (and very frequent) periods, which results in the variations in the geological record. Clube & Napier demonstrate that the breaking up of a giant comet produces a wide range of debris from objects 10 km across, to hundreds or thousands of 1 km-sized bodies, to multiple swarms of sub-kilometer-sized bodies. Many of these bodies have sooty, black surfaces making them almost impossible to see. Many of them are in an orbit very similar to the Taurid meteor streams, though a few may be in an orbit rotated about 90 degrees. Clube & Napier posit that many (if not most or all) of the asteroids in the solar system split from a giant comet (or many of them) thousands or tens of thousands of years ago, and it is the streams of debris that pose the most serious and immediate threats to our planet.

For example, one of the large asteroids in an earth-crossing orbit is named Hephaistos. It is about 10 km in diameter, about the same size as the asteroid that is depicted as striking the earth in the above-mentioned movie (the dinosaur extinction model). It is true that the effects of the impact of such a body would be felt globally, but it is not so clear that it would be exactly as 'global' as depicted in the movie. Nevertheless, the connection between a single impactor and past mass extinctions has been made and popularized widely, and this may be unfortunate considering the issues of more frequent and less global events that Clube & Napier address. The problem is, as they point out, a solitary large impact is, from an astronomical point of view, quite unlikely to be the only agency at work in such extinctions. Further, when one considers the details of the evidence, both astronomical and geological, many discrepancies in the single impactor scenario begin to emerge.

When the Alvarezes, *père et fils*, came across the iridium layer at the K-T extinction boundary, announcing that iridium in those amounts could only be thrown up by the impact of a large meteorite, this shocking idea was taken up gleefully by the press and everyone was on the hunt for iridium. Clube & Napier point out that there are several problems with the 'single impact' interpretation of the presence of iridium at the extinction boundary. The first problem is that *the concentration of the element is too high*. Why? Well, because if it were a single, giant impactor, such an asteroid would excavate several hundred times its own volume of earth crust material and blow it into the atmosphere mixed with its own material. This means that the iridium would be significantly diluted and would not precipitate on the planet in such concentrations as have been found. However, at many of the sites examined, it is noted that the iridium has been diluted by only 20 times its own volume (keeping in mind that the iridium in the comet/asteroid is already only a percentage of the total volume of the extraterrestrial body).

Additionally, other chemicals associated with the alleged single impact event do not fit the stony meteorite theory very well. There is an abundance of rare elements such as osmium and rhemium; enormous and overabundant common elements such as antimony and arsenic. In respect of this finding, Clube & Napier point out that, after a January 1983 eruption of Kilauea, particles collected from the volcano were found to have high concentrations of arsenic, selenium and other elements found in high abundance at the extinction boundary. These volcanic particles were also found to be very rich in iridium. Clube & Napier suggest that the iridium anomaly may, therefore, be a big red herring. They note: *"it is interesting to speculate whether, had a volcanic source of iridium been known in 1980, a meteorite impact would have been suggested"* by the Alvarezes? (Clube & Napier 1989, 225)

Probably not.

So, that was probably a good thing because it at least drew press attention to the matter since Clube & Napier also point out that there is an impressive amount of evidence that the extinction event was not just a process of evolutionary change and decay. Catastrophic changes – a profound ecological shock – took place across the Cretaceous-Tertiary boundary, and the devastation was certainly sudden. So the Alvarez theory opened the door to consider such possibilities in a world that was tightly bound up in uniformitarianism.

Among the interesting finds at this level of earth's history is that very large amounts of soot are also present at the extinction boundary. The conclusion is, of course, that global wildfires were raging during the extinction event. The movie tried to depict that with computer models (made on the assumption of a single large asteroid impact), which had the entire atmosphere of the earth heating up to the point where things just ignited spontaneously. That may not be exactly how things happen even with a very large meteor impact. Another point that Clube & Napier make is that there is not a trace of meteoritic debris in the form of stony inclusions in the sediments. I won't go into all the details; suffice it to say that it begins to look like the stray impact of a single 10 km-wide asteroid is not the cause of the global extinction after all.

What is a realistic scenario? Clube & Napier present the evidence that this extinction event was an episode of bombardment of many, dozens, hundreds, thousands of cometary fragment and/or meteorite-type bodies, some of them large, liberating copious amounts of meteorite dust in the earth's atmosphere, many of them exploding overhead in rains of fire. These swarms would be 'swimming' in streams of comet dust – tons of it – which would also be loading the atmosphere and precipitating onto the earth over months and years. The high concentrations of iridium found at the dinosaur extinction boundary at several localities, and the absence of bulk meteoritic debris, are hard to explain in terms of a single big bang but easily understood in terms of zodiacal dust as a provider of the input. Added to this, there is increasing evidence for a multiplicity of impacts at the dinosaur extinction boundary, as well as at other points of global catastrophe such as the Permian–Triassic (P–Tr) extinction event. The swarm theory also easily accounts for the huge amounts of soot at the boundary. An earth ablaze is within the capacity of an exceptionally intense swarm to produce, but probably beyond that of even a 10 km-wide single impactor. In short, the extinction of the dinosaurs may very well have been a complex, traumatic and prolonged affair.

Clube & Napier propose that the earth itself is a storehouse of information about its interactions with the galaxy, and that it is the galaxy itself, and earth's position in it, that drives the cycles of extinctions, mainly because the cycles of events best fit known galactic cycles.

The one thing that stands out from all of the evidence is the importance of very large comets that enter the solar system and break apart, leaving streams of debris that interact with our planet for millennia after the parent body or bodies have been captured and torn apart by intra-solar system forces. That such bombardments of the earth have occurred at other times is becoming more widely known; witness the work of Richard Firestone, Alan West and Simon Warwick-Smith who have identified the Carolina Bays as 'air impact' craters from overhead cometary explosions exactly like that of Tunguska. In fact, similar 'craters' were found in the Tunguska region with the exact same morphology. This event has been dated to about 12,500 years ago and was global in extent and cataclysmic in effect. Life on earth almost came to an end. What is frightening about this even is the sheer numbers of craters – upwards of 50,000 of them.

Clube & Napier mention the companion star hypothesis briefly, noting that *"Certainly the companion-star hypothesis adopts the central mechanism of the galactic one, namely the creation of comet showers through regular comet cloud disturbances.*" He then dismisses this as facing *"insuperable problems."* (Clube & Napier 1989, 233) The 'insuperable problems' are the proposed orbital periods for the hypothesized companion star and his idea that there would be far more cratering if the motive mechanism was a companion star. He may be entirely correct and his theory of galactic tides and comet birth in the cold, dark reaches of space certainly deals with the main elements of what we know about our celestial environment. As they note:

> The astronomical framework, grounded in celestial observations, is the basis for the theory of terrestrial catastrophism described here. ... it is in our view essential, if one is to arrive at a true picture, to take account of *all* the relevant evidence: 'hard evidence' in the geologist's sense has to be coupled with some respect for hard astronomical facts as well. Put another way, we do not need a 10-kilometre asteroid to land in our presence to demonstrate the amount of kinetic energy it will release. In particular, the correct picture must explain recent as well as past events in the terrestrial record. Thus the giant comet, and indeed the historical record, are essential elements in the quest for overall truth. It is this inextricable linkage between the very recent and the very remote past which lends urgency to the study: if we get the grand picture wrong, the next set of old bones in the ground could be ours. (Clube & Napier 1989, 235, 236)

We have presented some good evidence that Clube's and Napier's ideas are very likely correct or darn close: the earth has been repeatedly and regularly showered with extraterrestrial debris of some sort, and these showers have been generally disastrous from local scales, to regional, national, and even continental. It seems clear from the evidence that history itself is not a process of evolution, but more often, devolution, as each cosmic crisis has either resulted in 'survival of the lucky', as opposed to the fittest, and the more recent ones have been amplified or utilized by ruling elites to pursue their own agendas; or, on other occasions, the earth has suffered insults that have hardly turned a head in the human population. Tunguska was one such event.

On the night of 30 June and 1 July 1908, one of the most extraordinary events in modern history occurred. Tom Slemen writes in *Strange But True*:

> The first reports of a strange glow in the sky came from across Europe. Shortly after midnight on 1 July 1908, Londoners were intrigued to see a pink phosphorescent night sky over the capital. People who had retired awoke confused as the strange pink glow shone into their bedrooms. The same ruddy luminescence was reported over Belgium. The skies over Germany were curiously said to be bright green, while the heavens over Scotland were of an incredible intense whiteness which tricked the wildlife into believing it was dawn. Birdsong started and

cocks crowed – at two o'clock in the morning. The skies over Moscow were so bright, photographs were taken of the streets without using a magnesium flash. A captain on a ship on the River Volga said he could see vessels on the river two miles away by the uncanny astral light. One golf game in England almost went on until four in the morning under the nocturnal glow, and in the following week The Times of London was inundated with letters from readers from all over the United Kingdom to report the curious 'false dawn'. A woman in Huntingdon wrote that she had been able to read a book in her bedroom solely by the peculiar rosy light. There were hundreds of letters from people reporting identical lighting conditions that went on for weeks ... (Slemen 2000) [21]

None of the people witnessing this strange phenomenon had any idea that, in the central Siberian plateau, just after 7:15 a.m. local time, the planet had been hit by a cometary impactor that exploded – as most such impactors do – in the atmosphere just above the earth's surface. There was, of course, a great deal of comment about the strange, glowing sky in newspapers and scientific journals at the time. Clube and Napier write:

Some thought that icy particles had somehow formed high in the atmosphere and were reflecting sunlight. Others considered that a strange auroral disturbance was involved. The Danish astronomer Kohl drew attention to the fact that several very large meteors had recently been observed over Denmark and thought that comet dust in the high atmosphere might account for the phenomenon. But there was no agreement as to what had happened.

Over 500 miles to the south of the fall, a seismograph in the city of Irkutsk near Lake Baikal, close to the Mongolian border, registered strong earth tremors.

Nearly 400 miles south-west of the explosion, at 7:17 am on 30 June, a train driver on the Trans-Siberian express had to halt the train for fear of derailment due to the tremors and commotion. ...

In an Irkutsk newspaper dated 2 July it was reported that, in a village more than 200 miles from the Tunguska river, peasants had seen a fireball brighter than the sun approach the ground, followed by a huge cloud of black smoke, a forked tongue of flame and a loud crash as if from gunfire. 'All the villagers ran into the street in panic. The old women wept and everyone thought the end of the world was approaching.' (Clube & Napier 1989, 155–156)

In towns 300 to 400 miles away, hurricane-like gusts rattled doors, windows and crockery. This was followed within minutes by shockwaves that knocked down horses and hurled people working on boats into the river. The

---

[21] http://www.slemen.com/tunguska.html

closest observers of the explosion were reindeer herders asleep in their tents in several camps about 30 km from the site. They were blown into the air and knocked unconscious; one man blown into a tree later died of his injuries.

> Early in the morning when everyone was asleep in the tent, it was blown up in the air along with its occupants. Some lost consciousness. When they regained consciousness, they heard a great deal of noise and saw the forest burning around them, much of it devastated.
>
> The ground shook and incredibly prolonged roaring was heard. Everything round about was shrouded in smoke and fog from burning, falling trees. Eventually the noise died away and the wind dropped, but the forest went on burning. Many reindeer rushed away and were lost.[22]

Thousands of reindeer, in the general area around ground zero, were killed. Many campsites and storage huts belonging to the herders that dotted the area were destroyed. Clube and Napier continue:

> Local Siberian newspapers carried stories of a fireball in the sky, and a fearful explosion, but by the autumn of 1908 these stories had died out, and they went unnoticed in St. Petersburg (Leningrad), Moscow and the west. The region was arguably one of the most inaccessible places on Earth, in the centre of Siberia. ... However, rumours of an extraordinary event persisted, transmitted back by geologists and other intrepid researchers working in the area. These attracted the attention of a meteorite researcher, Leonard Kulik ... It was not until 1927 that an expedition ... led by Kulik, finally penetrated to the site of the 1908 explosion. (Clube & Napier 1989, 157)

Tom Slemen continues Kulik's story:

> Kulik got off the Trans-Siberian railway at the Taishet station and on horse-drawn sledges they set off on an arduous three-day odyssey through 350 miles of ice and snow until he and his men reached the village of Kezhma, situated on the River Angara. At the village Kulik and his party of researchers replenished their supplies of food, then struggled on for a three-day journey across wild and unchartered areas of Siberia until they reached the log-cabin village of Vanavara on 25 March.
>
> Kulik then tried to make headway through the untamed Siberian forests, or taiga as the Russians call it, but was forced to turn back after heavy snowdrifts almost froze the horses to death. For three days Kulik was forced to remain in the snow-

---

[22] http://earthsci.org/fossils/space/tunguska/tunguska.html

bound village of Vanavara, but during this period he interviewed many of the Evenki hunters who had witnessed the Siberian fireball's arrival on this planet.

The tales of the sky being ripped open by a falling sun and of a great thunder shaking the ground made Kulik even more eager to penetrate the taiga to find his holy grail.

When the weather gradually improved, Kulik set out for the Tunguska Valley. When he finally reached the site of the mysterious explosion, he was almost speechless. From a ridge overlooking the scene, Kulik took out his notebook and scribbled down his first impressions of the damage wreaked by the cosmic vandal. Kulik wrote:

"From our observation point no sign of forest can be seen, for everything has been devastated and burned, and around the edge of the dead area, the young, twenty-year-old forest growth has moved forward furiously, seeking sunshine and life. One has an uncanny feeling when one sees twenty to thirty-inch [thick] giant trees snapped across like twigs, and their tops hurled many yards away."

There were three further expeditions to the site of the Tunguska explosion, all of them headed by Kulik. In 1941, Hitler attacked Russia. The 58-year-old Leonid Kulik volunteered to defend Moscow, but was wounded by the Nazis. He was captured by German troops and thrown in a prison camp where he died from his wounds. (Slemen 2000)

The scientific investigation undertaken by Kulik in 1927 revealed that near the center of the blast many of the trees were still standing upright, even though denuded of limbs and leaves. Further from ground zero the trees were blown down and seared, forming concentric circles with the bases of the trees all pointing in the direction of the center of the blast. All of this evidence pointed to the fact that the blast almost certainly occurred in mid air. According to John Baxter and Thomas Atkins, in their book *The Fire Came By*, the explosion resulted in an enormous 'pillar of fire' and the blinding column was visible for hundreds of miles. The series of thunderous claps that followed could be heard for 500 miles or more.

Barograph recordings from around the world recorded the pressure waves of the blast, including stations between Cambridge, 50 miles north of London, and Petersfield, 55 miles south. Interestingly, it took the meteorologists in England twenty years to make the connection between their records and the devastation in Tunguska. As Clube and Napier note, the 'wave trains' were unlike any on record at the time, but nowadays we know that they do resemble those obtained from a hydrogen bomb explosion. From the available data (including the extent of flattened forest trees), the total energy released by the explosion equaled approximately 30-40 megatons, *"the combined force of a few dozen ordinary hydrogen bombs"* (Clube & Napier 1989, 158).

The noise of the explosion deafened those close to the blast. Following that, a searing hot thermal current from the fire in the sky raced across the forests. Tall conifers were scorched and ignited and the fires burned for days. Residents of Vanavara, a small trading post about forty miles distant, felt the fierce draft of heat. Some individuals there were flung into the air as the shock wave arrived; pieces of sod were gouged up, ceilings collapsed, and windows were shattered.

> *The date of fall (30 June) corresponds to the passage of the Earth through the maximum of the Beta Taurid stream.* From this and its trajectory, it appears that *the Tunguska object was part of the Taurid complex.* Probably the Earth passed through a swarm within the stream. The occurrence, this century, of an impact with the energy of a hydrogen bomb does give cause for some concern, and it is interesting to speculate on whether one's historical perceptions would be quite the same had the bolide struck an urban area or a city. (Clube & Napier 1989, 158–159)

Indeed, such an event would probably have triggered World War III. Just a few hours earlier or later and the impact could have been over a major urban area. But that didn't happen. As noted, it was twenty years before anyone really had an inkling of what *did* happen. That was partly due to the fact that the Russians in 1908 were somewhat occupied with politics. The previous year, 1907, Czar Nicholas had found himself faced with revolutionaries being elected in large numbers to the newly created parliament, the Duma. The eventual revolution began in 1917. One might even say that the Tunguska event was a harbinger of things to come, and maybe in more ways than anybody thinks. But even then, as Clube and Napier point out, the Tunguska impact was *"fairly trivial"*:

> 'In this year, on the Sunday before the Feast of St. John the Baptist, after sunset when the moon had first become visible a marvelous phenomenon was witnessed by some five or more men who were sitting there facing the moon. Now there was a bright new moon, and as usual in that phase its horns were tilted toward the east; and suddenly the upper horn split in two. From the midpoint of the division a flaming torch sprang up, spewing out, over a considerable distance, fire, hot coals, and sparks. Meanwhile the body of the moon which was below writhed, as it were, in anxiety, and, to put it in the words of those who reported it to me and saw it with their own eyes, the moon throbbed like a wounded snake. Afterwards it resumed its proper state. This phenomenon was repeated a dozen times or more, the flame assuming various twisting shapes at random and then returning to normal. Then after these transformations the moon from horn to horn, that is along its whole length, took on a blackish appearance. The pres-

ent writer was given this report by men who saw it with their own eyes, and are prepared to stake their honour on an oath that they have made no addition or falsification in the above narrative.'

This curious report is written in the chronicles of the medieval monk known as Gervase of Canterbury. The year of the event was AD 1178 and the date, 18 June on the Julian calendar, converts to the evening of *25 June* on the modern Gregorian one. If real, it is clear that some extraordinary event on the Moon is being described and the meteorite expert Hartung proposed that what was observed and recorded 800 years ago was the impact of a body on the Moon. The flame, he suggested, was the writhing of incandescent gases, or sunlight reflection from dust thrown out of the crater. The blackish appearance of the Moon along its whole length was a temporary suspension of dust buoyed up by a transient atmosphere.

... Hartung deduced that if there was a crater, it would be at least 7 miles in diameter, possess bright rays extending from it for at least seventy miles, and would lie between 30 and 60 degrees north, 75 and 105 degrees east on the Moon. ...

As it happens, there is one crater with the predicted characteristics exists. There is a crater named after the seventeenth-century heretic Giordano Bruno. This crater is located at 36 degrees N and 105 degrees E, within the predicted area. It is 13 miles in diameter and is distinguished by its remarkable brightness, and by the brilliant system of rays which extend several hundred miles out from it. [emphasis added] (Clube & Napier 1989, 159–160)

It should be noted that NASA has attempted to debunk Hartung's theory,[23] saying: *"Such an impact would have triggered a blizzard-like, week-long meteor storm on Earth – yet there are no accounts of such a storm in any known historical record, including the European, Chinese, Arabic, Japanese and Korean astronomical archives."* Well, we know from our current survey that this is not necessarily so. There could have been impacts on the earth that no one knew about – witness Tunguska – and it doesn't necessarily follow that an impactor on the moon would trigger a blizzard of meteors on earth.

Eighty-six years after Tunguska, in July of 1994, there was another harbinger: the fragments of Comet Shoemaker-Levy struck Jupiter. If they were able to brush Tunguska off as a fluke, scientists were not so easily able to put aside the spectacle of a whole string of comets hitting another planet in our solar system, one after the other, as the planet spun in space. In the same year, a book entitled *Hazards due to Comets and Asteroids* was published in reaction to this, then impending, event. The book is a collection of papers, one of which says:

---

[23] http://science.nasa.gov/science-news/science-at-nasa/2001/ast26apr_1/

Our understanding of the history of Earth and its inhabitants is undergoing a radical change. The gradual processes of geologic change and evolution, it is now clear, are punctuated by natural catastrophes on a colossal scale – catastrophes resulting from collisions of large asteroids and comets with Earth. It is, to use the popular term, a "paradigm shift."

This "new catastrophism," is not unlike the revolutions brought about by the heliocentric solar system of Copernicus, or Darwinian evolution, or the big bang. In retrospect, such revolutionary ideas always seem obvious. On reading the Origin of Species, Thomas Huxley remarked simply: "Why didn't I think of that." Now, looking at the Moon, we find ourselves wondering why it took so long to ask whether the process that cratered its surface is still going on. (Park, Garver and Dawson in Gehrels 1994)

Let me repeat that most important remark:

Our understanding of the history of Earth and its inhabitants is undergoing a radical change. The gradual processes of geologic change and evolution, it is now clear, are punctuated by natural catastrophes *on a colossal scale* ... [emphasis added]

That may be the understatement of the millennium.

It has been suggested by a number of researchers mentioned in this book that the current 'climate change' issues are actually due to the earth moving through cosmic dust clouds and that all the hoopla about global warming is simply a cover-up of this fact. (It could even be that such things as 'chemtrails' are a result of such dust loading in the upper atmosphere). Clube suggests that it is a cometary cosmic dust cloud, left over from the break-up of a giant comet that, for a long period of the earth's history, threatened and bombarded our planet with unspeakably horrible, civilization-destroying fragments, i.e., the progenitor of the Taurid stream, including the Tunguska comet. Clube also argues that these events were the basis for the formulation of humanity's ideas about the cosmos, gods, religion and even astrology. Over time, as the giant comet spent most of its mass in its Titanic fury, dying away to occasional less-than-civilization-destroying bombardments, our conceptions of gods changed; the reality was tossed out the window in favor of fairy tales for both science and religion, not to mention astrology. He and Napier write:

Three thousand years ago, in accordance with age-old practice, the kings of Babylon were still employing astronomer-priests to give warnings of cosmic visitations. A thousand years ago, the emperors of China were still relying on similar skills, while in Europe the Pope saw messages in the sky and urged Holy War. But

113

this latter was an aberration; for *the last two and a half thousand years have seen the decline and fall of the sky gods, and the growing presumption that the cosmos is stable and regular. The shift of paradigm has been unconscious, convenient, insidious and thorough.* Probably, the rediscovery of a lost tradition of celestial catastrophe could not have been made through analysis of ancient texts alone; a key had to be provided, and it has been, by the paraphernalia of modern science. It is a salutary lesson both on the capacity of human reasoning to get it wrong for long periods of time, and on the essential unity of knowledge.

It would be naïve to think, however, that one merely has to point to deep-seated cracks in the structure of modern knowledge to have scholars setting to and constructing a better framework within which mankind might plan his future. There is considerable intellectual capital invested in the status quo, enough to ensure that *those with an interest in preserving it,* the 'enlightened' and the 'established', will continue to present the cosmos to us in a suitably non-violent form. The history of ideas reveals that some will even go further *and act as a kind of thought police, whipping potential deviants into line. For them, temporal power takes precedence over the fate of the species.* [emphasis added] (Clube & Napier 1989, 276–277)

This problem of the 'status quo' and 'thought police' is not a minor issue. In this book, we have repeatedly come face to face with the obvious fact that those in positions of power and authority lie to the masses of humanity as a rule rather than an exception.

Again and again, we have discussed historical cometary bombardments, the consequences for humanity being dire enough to begin with, but which were then exacerbated and capitalized upon by pathological deviants, following which the facts were covered up by lies. Worse, again and again we have seen that the masses of humanity that suffer the most from these assaults and manipulations appear to be quite willing to be deceived, even to their own death and destruction (if only someone will give them something warm and fuzzy to believe in and a scapegoat to blame). At present, we are observing this phenomenon playing out on the global stage in real time and again and again we shake our heads and ask 'why?' What is it about our world, our present day culture, and the human beings that occupy our planet, that gives rise to this bizarre condition that lies are preferred over truth, death over life?

As I have pulled on the many threads that lead to and away from these matters, what I see repeatedly is a small group of people on the planet who rule over the masses of people, who do not have humanity's best interests at heart – another staggering understatement. What I see is the constant – and mostly successful – efforts of this small group to enforce totalitarianism in one guise or another – from politics to religion to all fields of science – on the entire world. And again and again they utilize disasters as a means to consolidate their power. Read Naomi Klein's book *The Shock Doctrine* to get a good grip on exactly how this works now, and how it has always worked. There is nothing new under the sun.

Parallel to the growing awareness of pathology in power is the expanding research among a few maverick scientists and researchers showing plainly that cometary disasters are cyclic and it is altogether possible that there are still a few big bangs left in the Taurid meteor stream. It is also possible that there are new swarms of comets heading our way as recent 'global warming' and 'moon capture' events on the other planets seem to indicate. Something is definitely going on in our solar system and we need to know what it is. What seems certain is that if disasters are in our future, it will be seen by the elite as just another opportunity to use their 'shock' tactics to consolidate their power over the entire globe, never mind that there might be so few people left that it will only amount to being a big frog in a small pond – and possibly a frozen one at that.

In her book *The Origins of Totalitarianism*, Hannah Arendt wrestles mightily with the problem, never quite reaching a complete explanation or solution. After all she saw, all she experienced, all her research, she wrote in the introduction:

> Two world wars in one generation, separated by an uninterrupted chain of local wars and revolutions, followed by no peace treaty for the vanquished and no respite for the victor, have ended in the anticipation of a third World War between the two remaining world powers. This moment of anticipation is like the calm that settles after all hopes have died. We no longer hope for an eventual restoration of the old world order with all its traditions, or for the reintegration of the masses of five continents who have been thrown into a chaos produced by the violence of wars and revolutions and the growing decay of all that has still been spared. Under the most diverse conditions and disparate circumstances, we watch the development of the same phenomena – homelessness on an unprecedented scale, rootlessness to an unprecedented depth.
>
> *Never has our future been more unpredictable, never have we depended so much on political forces that cannot be trusted* to follow the rules of common sense and self-interest – *forces that look like sheer insanity*, if judged by the standards of other centuries. *It is as though mankind had divided itself* between those who believe in human omnipotence (who think that everything is possible if one knows how to organize masses for it) and those for whom powerlessness has become the major experience of their lives.
>
> On the level of historical insight and political thought there prevails an ill-defined, general agreement that *the essential structure of all civilizations is at the breaking point.* Although it may seem better preserved in some parts of the world

115

than in others, it can nowhere provide the guidance to the possibilities of the century, or an adequate response to its horrors. Desperate hope and desperate fear often seem closer to the center of such events than balanced judgment and measured insight. The central events of our time are not less effectively forgotten by those committed to a belief in an unavoidable doom, than by those who have given themselves up to reckless optimism.

This book ... was written out of the conviction that *it should be possible to discover the hidden mechanics* by which all traditional elements of our political and spiritual world were dissolved into a conglomeration where *everything seems to have lost specific value, and has become unrecognizable for human comprehension, unusable for human purpose.* To yield to the mere process of disintegration has become an irresistible temptation, not only because it has assumed the spurious grandeur of "historical necessity," but also because everything outside it has begun to appear lifeless, bloodless, meaningless, and unreal.

The conviction that everything that happens on earth must be comprehensible to man can lead to interpreting history by commonplaces. *Comprehension does not mean denying the outrageous, deducing the unprecedented from precedents, or explaining phenomena by such analogies and generalities that the impact of reality and the shock of experience are no longer felt.* It means, rather, examining and bearing consciously the burden which our century has placed on us – neither denying its existence nor submitting meekly to its weight. Comprehension, in short, means the unpremeditated attentive facing up to ... reality ...

In this sense, it must be possible to face and understand the outrageous fact that so small (and, in world politics, so unimportant) a phenomenon as the Jewish question and anti-Semitism could become the catalytic agent for first, the Nazi movement, then a world war, and finally the establishment of death factories. ... *or the curious contradiction between the totalitarian movements' avowed cynical "realism" and their conspicuous disdain of the whole texture of reality.* ...

The *totalitarian attempt at global conquest and total domination* has been the destructive way out of all impasses. *Its victory may coincide with the destruction of humanity; wherever it has ruled, it has begun to destroy the essence of man.* (Arendt 1951)

Hannah's comments and observations of our world, wrung from her own pain and experiences, have never been more poignant than today when we face exactly what she was describing: global totalitarianism rising like a mighty juggernaut with the end of humanity in sight. And never have we seen more clearly that characteristic of the authoritarian type Hannah also saw: *"avowed cynical 'realism' and ... conspicuous disdain of the whole texture of reality."* If you want to see a stunning portrayal of this type of individual, get a copy of *Lions for Lambs* and observe the character played by Tom Cruise.

On my news analysis website, SOTT.net, we regularly discuss the problem of pathological deviance and how pathology can drive an individual to seek power over others. We have brought forward the work of Andrzej Lobaczewski, *Political Ponerology*, and this goes a long way toward providing a framework in which the history of evil – particularly political evil – can be understood.

As far as I can tell, Hannah Arendt did not consider the problems of pathology and how it operates in society as a corrupting element, nor did she consider the factor that Judaism and its offspring, Christianity and Islam, could be the carriers of the disease of totalitarianism. In this sense, it is really important to come to the knowledge of how religions are created and by whom – generally pathological deviants – and how they are used, in combination with politics, to control masses of people. This, of course, takes us back to the problem of comets in our skies and impacting our planet.

As it happens, after pulling on so many threads relating to the topic, it occurs to me that comets may, indeed, have a great deal to do with the major social problem on our planet today: psychopaths.

One might reasonably ask: Is some evolutionary process at work here? Clearly, staggeringly large numbers of people die repeatedly when they place their trust in lies and liars. And nearly as often do the liars in power find themselves in difficult situations as a result of their over-reaching and ignoring facts. Obviously, if evolution is at work here, those individuals – and their offspring who believe lies – are ultimately eliminated from the gene pool. What happens to those who, as Arendt suggests, try to comprehend, understand, and face the facts of our reality, remains to be seen. As she also states, this comprehension must not deny the outrageous, nor attempt to deduce the unprecedented from precedent.

You see, evolutionarily speaking, psychopaths should not exist. Throughout history it can be seen that human beings have needed to co-operate and care about one another in order to survive and produce a new generation that will carry on the processes of society. Most human dynamics are based on people trying to work out their problems and come to resolutions agreeable to the greatest number or, at the very least, in the interactions between two people. The issue of trust is paramount. Someone who betrays your trust is someone you cannot live or work with. Therefore, psychopaths, who are untrustworthy, should have long ago become extinct. But that isn't the way things are. It appears, in fact, as if psychopathy has increased.

As we can observe throughout history right down to the present day, being the only psychopath in a group of trusting people can be a very good thing for the psychopath. As groups get larger, they can accommodate more psychopaths. It seems that when the number of people carrying the 'psychopathy gene' is small enough, those few who carry it can achieve phenomenal breeding success. As Glenn Whitman explains it in his article 'Psychopaths as Hawk Strategists':

> What's nice about this explanation is that it not only explains why psychopaths exist, but also why we're not all psychopaths. If there are few enough psychopaths in the population, then being a psychopath makes sense because you'll mostly have winning confrontations with nice people. But if there are too many psychopaths, then the gains from taking advantage of nice people will be swamped by the losses from confronting other psychopaths. In equilibrium, you'll get both psychos and nice folks, with each strategy generating approximately equal returns, and with the precise balance determined by the relative payoffs of different interactions.[24]

The problem is, as noted, we are living in a time when psychopathy seems to have increased almost exponentially. Moreover, as Hannah Arendt notes – and this was never truer than today – the broad sweep of history indicates that the psychopaths are winning and that means destruction for all, including the psychopaths.

Which leads us to the problem: psychopathy being what it is and doing what it does, should certainly have brought the human race to total destruction a long time ago *if*, as a taxon, it had existed throughout the development of humankind. And that suggests that it did not. In fact, the studies of Marija Gimbutas indicate that there was a time when psychopathy was not 'at the top'.

As far back as we can go with archaeological records, i.e. 'hard science,' we find that the worship of the Great Celestial Goddess was the act of veneration of the Universe and all within it as the living body of the Goddess-Mother-Creatrix. This goddess was symbolized by the double wavy lines of water – both the cosmic sea of potential/activation, as well as the life-giving moisture that emanated from the body of the goddess and all women as her representatives on earth; the oceans and seas, rivers and springs and wells.

She was the sky goddess and the earth was her womb and all upon it were her children. The 'Son/Sun-King' died every year and was reborn in the passing of the seasons. Rites and rituals were enacted to insure the rebirth of the 'Son-Sun' through the 'womb of the Earth', the generative organ of the cosmic Mother. There was a purity and innocence … a pastoral, Arcadian simplicity and symmetry to life. Men and women were equals in importance as the 'twin offspring' of the goddess.

Women were honored and cherished in their three manifestations: the virgin-future-mother; the mother-nourisher; and the crone-mother of wis-

---

[24] http://agoraphilia.blogspot.com/2007/11/psychopaths-as-hawk-strategists.html

dom. Men were partners and protectors of women, thereby protectors of their own being since all were born and nursed by women. The male energy served the female because the female served the male; there was cycling energy, synergy, symmetry and balance.

The wavy lines of water, the Cosmic Sea of the Mother, were, at some point, corrupted into the symbol of the serpent; the woman became associated with the serpent instead of the stars; and everything went downhill from there. This event is described in Genesis 3:19 where Adam is told by Yahweh: *"In the sweat of your face you shall eat bread till you return to the ground, for out of it you were taken; you are dust, and to dust you shall return."*

Here, the 'dust' or the earth is denigrated and reviled as worthless in the same passage as woman herself is denigrated and reviled as the temptress. But, in the pre-existing goddess religions, the earth out of which all life has been born is *not* dust, but alive – as the womb of the goddess herself. And, judging by the massive evidence collected by Marija Gimbutas, this was the most ancient and worldwide order of thought, antecedent to and underlying all other myths, predating the male-dominated pantheons by many thousands of years.

> The main theme of the Goddess symbolism is the mystery of birth and death and the renewal of life, not only human, but all life on earth and indeed in the whole cosmos. Symbols and images cluster around the parthenogenetic (self-generating) Goddess and her basic functions as Giver of Life, Wielder of Death, and, not less importantly, as Regeneratrix, and around the Earth Mother, the Fertility Goddess young and old, rising and dying with plant life. She was the single source of all life and who took her energy from the springs and wells, from the sun, moon, and moist earth. This symbolic system represents cyclical, not linear, mythical time. In art this is manifested by the signs of dynamic motion: whirling and twisting spirals … (Gimbutas 1989, xix)

At some point the goddess, or source of moisture and all life, was identified with the serpent-as-tempter and became the wicked temptress herself. There is some difficulty in untangling the threads of the currently known symbolic systems of which there are now two: one reflecting a matriarchal serpent culture, and the other an androcratic, militaristic serpent culture. The man and woman, who were formerly united as the original 'Twins of Creation', and who, in their union could 'control' the serpent, became opposed to each other: unbalanced, antagonists, not partners. The benevolent 'Son-King', the leader of the flock, was no longer sustained and supported by the female energy, channeled from the goddess through the 'Tree,' in the body of the goddess' representative on earth, Woman, but instead became the male 'killer' of the serpent *and* controller/oppressor of the goddess, who became the exemplification of temptation and occasional consort of the serpent.

At some point, the ancients say that Eve betrayed Adam, Cain killed his brother, Abel, and psychopathy was let loose on our planet. Who was this serpent? What are the clues that point to his appearance on the scene? Where? When? Tunguska was the first key to this problem.

If it had not been for the cometary impact at Tunguska in 1908, and then the Shoemaker-Levy impacts on Jupiter eighty-six years later, our attention would not be drawn to certain aspects of our history that have been thoroughly covered up, nor the effects of those lies and distortions on our world today. Somehow, psychopaths and comets are inextricably bound together; it is even possible that the same cosmic forces that sent comets our way also 'created' psychopaths. This is what I want to explore in the final chapters of this book.

Our next stop, then, is the end of the last Ice Age when the Cosmic Serpent appeared on the scene and the processes that have led to our involvement in the Sixth Extinction got underway. But first, I want to round out our little study of history with something a little more comprehensive.

PART II:

THE LIST

8. Meteorites, Asteroids, and Comets: Damages, Disasters, Injuries, Deaths, and Very Close Calls

Astronomy books and papers far too numerous to cite offer the assurance that "no one has ever been killed by a meteorite."
– John S. Lewis, University of Arizona.

I have already mentioned the book *Hazards Due to Comets and Asteroids* edited by Tom Gehrels, with 120 contributing authors, published by the University of Arizona Press in 1994. As noted, it was published in reaction to the impending Comet Shoemaker-Levy event vis-à-vis Jupiter. It contains a paper beginning on page 1225 that is written by Robert L. Park of the American Physical Society, Lori B. Garver of the National Space Society and Terry Dawson, a staffer for the House Committee on Science, Technology and Space working for the Committee's then Chairman, Rep. George Brown. The following is a condensation of the main points of this paper:

Our understanding of the history of Earth and its inhabitants is undergoing a radical change. The gradual processes of geologic change and evolution, it is now clear, are punctuated by natural catastrophes on a colossal scale – catastrophes resulting from collisions of large asteroids and comets with Earth. It is, to use the popular term, a "paradigm shift."

This "new catastrophism," is not unlike the revolutions brought about by the heliocentric solar system of Copernicus, or Darwinian evolution, or the big bang. In retrospect, such revolutionary ideas always seem obvious. On reading *The Origin of Species*, Thomas Huxley remarked simply: "Why didn't I think of that." Now, looking at the Moon, we find ourselves wondering why it took so long to ask whether the process that cratered its surface is still going on. ...

The long time scale between major impacts has implications for public policy. Governments do not function on geologic time. On the North Dakota prairie near the town of Grand Forks, lie the abandoned ruins of America's ballistic missile defense system. ... Built in accordance with the ABM treaty, the Grand Forks facility was meant to defend our retaliatory capacity. It was declared operational in 1975 – and decommissioned the same year. National leaders had been persuaded by some scientists that the Grand Forks facility would meet the threat to our intercontinental ballistic missile fleet, even though other scientists warned that the system was dangerous and ineffective. It was closed because the money to operate it was needed for other projects that were deemed to be more urgent.

The lesson of Grand Forks is as old as human history: societies will not sustain indefinitely a defense against an infrequent and unpredictable threat. Governments often respond quickly to a crisis, but are less well suited to remaining prepared for extended periods. Even on the brief scale of human lifetimes, resources are eventually diverted to more immediate problems, or defenses are allowed to decay into a state of unreadiness. According to news accounts, in the great flood of 1993, the U.S. Corp of Engineers prepared to close the massive iron gates in the vast complex of levees on the Mississippi and its tributaries only to discover that some of the gates had been removed and sold for scrap. Periodic inspections had been suspended to save money. Indeed, civilization will do well to survive long enough to be threatened by a major asteroid impact; our own destructive impulses of the unanticipated consequences of our technologies seem likely to do us in first. It is unrealistic to expect governments to sustain a commitment to protection against a rare occurrence when they are constantly under pressure to respond to some perceived immediate crisis.

Particularly now [1994], with nuclear weapons being dismantled by the major powers, any talk of a nuclear defense against such an unlikely hazard as cosmic collisions will be seen as an effort by the weapons community to sustain itself. *The risk of diversion of any mitigation system to military uses must be regarded as a more immediate hazard.* ...

Given the frequency of past collisions, major impact is unlikely to occur in the next century. ...

Discussion of mitigation may serve one public purpose. It is important that devastation not be accepted as inevitable, otherwise society might prefer not to know when it is coming. An asteroid interception workshop hosted by NASA in 1992 concluded that available technology can deal effectively with a threatening asteroid, given warning time on the order of several years. That conclusion validates the view that current efforts should concentrate on detection and orbit determination.

The challenge of science is to identify objects that threaten Earth and work out the timetables for their arrival. Here the challenge is straightforward and technical. ...

The emphasis has properly been on impacts that would be expected to have global consequences. Even for objects too small to produce more than local effects, however, it has been pointed out that an impact might be misidentified as a nuclear explosion. Misidentification would be most likely among nations that have recently joined the ranks of "nuclear powers" and would therefore be expected to have less sophisticated means of verification.

It is more than a hypothetical concern. We recall that the 1978 South Indian Ocean anomaly, detected by a Vela satellite, was suspected at the time of being a South African-Israeli nuclear test. In spite of the failure to find any confirming evidence from intelligence sources or atmospheric monitoring, it created international tensions that lasted for years.

At the time, there were suggestions that it might have been an artifact produced by micrometeorite impact on the Vela satellite itself, but little serious consideration seems to have been given to the idea that the satellite had observed the fireball from an asteroid impact in the atmosphere. A 1990 satellite observation of an apparent asteroid impact fireball over the Western Pacific has been described by Reynolds (1993). The danger of misidentification, which grows as weapons proliferate among less sophisticated nations, is meliorated in part by publicizing the possibility. The only sure means of avoiding an unfortunate response, however, would be for everyone to know the impact is coming. Which again places the emphasis on detection.

Efforts to persuade governments to invest significant resources in evaluation of the hazard of asteroid impacts must overcome what has been called the "giggle factor." Clearly, elected officials in Washington are not being inundated with mail from constituents complaining that a member of their family has just been killed or their property destroyed by a marauding asteroid. ...

Congressional involvement has been confined to the Committee on Science, Space and Technology of the U.S. House of Representatives, whose current chair, George Brown of California, has maintained an interest in the asteroid issue for several years. The committee directed NASA to conduct two international workshops on the asteroid threat. ...

In March of 1993, the Space Subcommittee held a formal hearing to examine the results of the two workshops. Some members remain skeptical that the threat is real. But even among those who recognize that it is only a question of when a major impact will occur, there was no sense of urgency. ...

The frequency of impacts of objects of various sizes is known only to limited precisions. In particular, objects up to several meters in diameter explode in the atmosphere without reaching the surface. Although the energy released in these explosions may be many times greater than that released by the Hiroshima bomb, they most frequently occur over the ocean or sparsely inhabited regions of Earth and go unreported. ...

Congress is unlikely to take any action in the absence of public pressure. Once the public understands that Earth and the life on it have been shaped by cosmic collisions (and the process is continuing), they will be more likely to support the science needed to evaluate the threat. The scientific community must, therefore, concentrate on public education. ...

All of this creates a dilemma. While it is important to inform the public, it is dangerous to encourage fear mongering. ... Scientists would do well, for example, to avoid such terms as "near miss." The public understands "near-miss" as the draft of wind from a truck that passes as you step off the curb – not a truck that went by six hours earlier. ...

Even in such staid newspapers as the *New York Times* and *Washington Post*, articles may include a well-reasoned discussion of relative risk, but the headline writers find "doomsday rock," "space bullets" and "killer comet" irresistible. These headlines exploit the excessive fear engendered by events people feel powerless to control. The image of an indifferent mountain of stone and metal guided by the immutable laws of physics toward an inevitable rendezvous with Earth, is the stuff of nightmares. Remarkably, however, *Nature* has apparently provided a non-threatening demonstration. The impact of comet Shoemaker-Levy 9 on the back side of Jupiter in July of 1994 provides an historic opportunity to educate the public without terrorizing anyone.

Shoemaker-Levy 9, in its last pass by Jupiter, broke into a string of 21 major pieces. The energy released by the impacts of the full string will be equivalent to about a billion megatons of TNT. Although the pieces will impact on the side of Jupiter away from Earth, millions of amateur astronomers will be watching to see the flashes reflected from Jupiter's moons. A few hours later, the rotation of Jupiter will bring the impact region into view. There is great disagreement about what will be seen, but no one suggests that it will not be spectacular.

The asteroid-comet community needs only to insure that everything is fully and accurately explained; the message will take care of itself: (1) the energy deposited by the cosmic impacts is enormous (2) this is a process that is still going on.

The authors clearly had a lot of faith in human beings and thought that all scientists had to do was to tell the public the truth and they would get enough support to fund cataloging the dangerous asteroids in earth-crossing orbits. They also thought that this was the main problem: asteroids that could be seen and listed.

What seems obvious to me is that someone else took the 'Lesson of the Grand Forks Facility' in an entirely different way. The question that comes to my mind is this: are the elite powers creating a War On Terror as an immediate and constant pressure on the public to get the needed support for the stockpiling of nuclear weapons so they will have them to use on asteroids? You know, a benevolent lie with a million plus innocent Iraqis being sacrificed to sustain it. Kind of like Madeleine Albright: In 1996 then-UN Ambassador Albright was asked by *60 Minutes* correspondent Lesley Stahl, in reference to years of U.S.-led economic sanctions against Iraq, *"We have heard that half a million children have died. I mean, that is more children than died in Hiroshima. And, you know, is the price worth it?"* To which Ambassador Albright responded, *"I think that is a very hard choice, but the price, we think, the price is worth it."* So, is there somebody at the top who thinks that stockpiling nuclear weapons is a good thing for planetary defense of a cosmic nature?

There is another way to ask the question: are the powers that be using the threat of asteroids on lawmakers to get them to agree to backing the phony War on Terror in order to obtain and retain the support of the masses when what they are really doing is just planning on a fascist takeover of the world? Notice that the paper above also said: *"The risk of diversion of any mitigation system to military uses must be regarded as a more immediate hazard."*

It's hard to tell what goes on in the minds of deviants. One thing I think we can be sure of is that the threat of cometary bombardment is real and immediate, and that comes from the science. Sadly, it does not come from our leaders who, even if they are aware of some threat and are stockpiling nuclear weapons to use to divert inbound asteroids or comets, haven't bothered to make the threat clear to the masses of humanity via science, as they very well could.

Scanning through this almost 1300-page volume, which collects pretty much all the then-scientifically acknowledged data on comet and asteroid impacts, reveals that there was some pretty interesting thinking going on prior to Shoemaker-Levy 9. We've come a long way in our understanding since then; well, some have. The U.S. school is still pretty much stuck in the 'single massive asteroid at vast timescales', probably due to political pressures to keep the real issues covered up. I noted that Shoemaker had a paper in the volume where he said there were only 140 known impact craters on the earth. He completely ignored the 50,000 Carolina bays, which have been reclaimed for what they are by Richard Firestone, Allen West and Simon Warwick-Smith in *The Cycle of Cosmic Catastrophes*.

We also note the remark in the above paper: *"The frequency of impacts of objects of various sizes is known only to limited precisions. In particular, objects up to several meters in diameter explode in the atmosphere without reaching the surface."*

Obviously, this guy wasn't part of the same crowd that hung out with Brigadier General S. Pete Worden, who said that he believes *"we should pay more attention to the 'Tunguska-class' objects – 100 meter or so objects which can strike up to several times per century with the destructiveness of a nuclear weapon"*, reported in the previous chapter. In any event, the authors of the above quoted paper had a generally open attitude toward the public and educating them that no longer seems to be the perception of our ruling elites.

Over the past few years, while our research team has been tracking the increasing flux of fireballs and meteorites entering the earth's atmosphere, we have been, by turns, amused and horrified at the ignorant reactions and declarations that issue from academia and the media regarding these incursions. A few years ago, we read that 'this is a "once in a hundred years" event.' Not long after it was a 'once in a lifetime' event. Still later, after many more incidents it became a 'once in a decade' event. More recently, it has been admitted in some quarters that meteorites hit the ground (as opposed to safely burning up in the atmosphere) several times a year.

We have discovered the fact that the governments of our planet are well aware that there are atmospheric explosions from such bodies numerous times a year. We have also learned that the frequent reports of unusual booms and shaking of the ground is often due to such overhead explosions. Yet the media steadfastly refuses to honestly address this issue, offering instead a plethora of articles presenting opposing academic arguments designed to put the populace back to sleep, reassuring them that there is nothing to worry about, that such things only happen every 100,000 years or so, and certainly, the Space Watch Program is going to find all the possible impactors and take care of things. We most certainly can see that the issue of meteorite, cometary and asteroid impacts on our planet, and their true potential danger to each and every one of us, must be added to the list of unfunded and scientifically verboten research.

This is a very bad and dangerous state of affairs.

Considering the nature of the topic and the obvious efforts to marginalize it, to cover it up and transfer it to the realm of 'crazy conspiracy theories' or worse, we had some concern about Professor Victor Clube, co-author of *The Cosmic Winter* and the paper addressed to the European Office of Aerospace Research and development, dated 4 June 1996, entitled 'The Hazard to Civilization from Fireballs and Comets'. After a little research however, it turned out that Victor Clube had retired. One of our researchers who is a climate scientist at a major U.S. research facility, undertook to try to find Clube and eventually received a faxed letter from the good professor which I am quoting here in part (my italics):

> 2008 Feb 15
> […]
> I note that my health is questioned and I hasten to admit its comparative rudeness! … In fact I still like to think my apparent inactivity is not quite as total as a google search might be indicating!
> First, I should say your references to the (cosmically complacent) paleoclimate community and to my otherwise unread narrative report to the USAF European

office strike a very considerable chord with me. After all neither Ms Victoria Cox nor your good self can be aware how very much Bill and I had reason to appreciate the timely injection of USAF funds at a time when *the line of research we championed appeared to be successfully closed down by the UK scientific establishment.* Thus *we were both in turn obliged to relinquish our career posts at the Royal Observatory, Edinburgh on account of this line of research* – which gave rise to our reincarnation at a more tolerant haven namely my alma mater (Oxford).

Also, whilst I broadly accept your commentary regarding the role of "national elites" in the face of near-Earth threats, *I am quite certain the elites in practice currently know* VERY *"much* LESS *than they let on"* and that the situation for humanity is dire. Any comfort you may draw from the opposite opinion seems to me to be entirely misplaced. Thus although the globally modest efforts to assess the NEO threat with telescopes by a few semi-enlightened national administrations (e.g., USA) or by a few private enterprises (e.g. Gates) are certainly to be commended, I look upon this aspect of the NEO threat as basically intermittent and therefore more or less symbolic so far *as generally more urgent and still largely undetected low mass NEO flux (which is demonstrably climatological in its effect)* is concerned. This particular threat (evidently responsible for our planet's evolving glacial/interglacial condition during the past 3 million years) is of course <u>fundamentally</u> ignored by the current Body Scientific and hence by most of humanity as well.

Why is the Body Scientific so misguided? Basically, in my view, because many these days are unaware <u>two</u> secular versions of natural philosophy have arisen since the West's renaissance when Plato's works were reintroduced, essentially substantiating a zodiacal circulation of gods apparently ancestral to the European elite.

Following the invention of telescopes and the undigested revelation that zodiacal gods appeared non-existent (Galileo), the elite began to invoke transcendental divinity whilst rejecting any material circulation or "unmoved mover" occupying interplanetary space. Newton however delved behind Plato to reach Pythagoras etc. and revived the claims for a material circulation in the zodiac (diminished in luminosity) whose encounters with our planet remained a source of recurring providence.

With advancing technology and the improvement of telescopes, the zodiac then revealed the first near-Earth comet (P/Encke) and the first not-so-near asteroid (Ceres) thus re-opening the question whether the ancient zodiacal circulation comprised near-Earth comets.

An ensuing but very severe crisis then emerged within the Western elite which was resolved ca 1830 in favour of the preferred secular version of natural philosophy (Galileo's) as opposed to the other (Newton's).

I now believe this turn of events was clearly initiated by the Royal Astronomical Society's official charter in Britain (top nation!) as soon as it agreed <u>not</u> to publish observations of P/Encke even though its namesake had already received the Society's highest commendation.

Arising from the political interference, Newton's unpublished papers were to remain concealed for at least another century whilst climatology (like many other branches of knowledge) was never properly integrated with Newtonian astrophysics and geophysics.

Once this thread of advancing knowledge is fully appreciated this integration may be reckoned to have only recently been attempted with the publication of an excellent interdisciplinary monograph "Ice Ages and Astronomical Causes" (Springer 2000) by Muller and McDonald (deceased). This seminal climatological work once again connects climate with a material circulation in the zodiac and can indeed hardly be faulted until it reaches its chapter 7 (Accretion climate models). Here, unfortunately, it becomes heavily dependent on a completely mistaken understanding of previously unknown dust bands in the zodiac revealed by IRAS as recently as 1983-4*.

The IRAS team was a typical post-WWII scientific enterprise lasting only a decade or so and lacking a secure institutional base but inspired and funded by the top nation (USA) for the benefit of humanity. Its claims under these conditions, like those for resurrection, were endorsed far too rapidly by the Body Scientific and it was absolutely wrong to suppose only low eccentricity (orbital) material could be invoked in creating these previously undetected dust bands. The IRAS team thus completely failed to associate these bands with their most obvious source namely the long term progenitor of Comet P/Encke.

The upshot of all this is a succession of UK and US astronomical/cosmological elites skillfully distracting the Body Scientific and humanity for the past 200 years from the most prominent material dominating near-Earth space and controlling Earth's climate for the past 3 million years (and a no doubt comparable period in future). The social/political outcome of all this (including the current global warming scare) could be laughable if it were not also so deep and profound. [...]

*P. S. The authors of these publications both fail to realise the conclusion between our planet's climate and it's spin-orbit differential nodal precession must arise on account of its long term orbital resonance with the accretion source (or the ancient music of the spheres). As you know, our research builds upon Comet P/Encke's proximity to such a resonance.

As I have mentioned already, everyone needs to read Clube's books – *The Cosmic Winter and The Cosmic Serpent* – if you can possibly get copies. They are out of print and you may have to go to a library to do so. The reason these books are important is because they give you a good idea, in very realistic terms, of what you may have to deal with at some point in the not-too distant future ... *and*, that it is eminently survivable *if* you know what to look for and how to prepare.

We aren't talking about a giant asteroid here that is going to create a global wave of firestorm destruction like the movies depict. It's not 'Planet X' or 'Nibiru'. It's not necessarily the 'End of the World'. However, it might be somewhat like the Black Death with the loss of half the population of one or more continents, or the great Chicago Fire or Tunguska, or even all of these rolled into one; but all of those events were survivable had the victims been informed and prepared. Yes, there are those who, even had they been informed and prepared, would not have survived in any case, but we choose the optimistic path: Knowledge Protects. And we are trying our hardest to give you that knowledge.

The very fact that Bill Gates and others have invested in a seed bank, in an observatory, suggest to us strongly that they (the rich and powerful), too, have this sort of future in mind. The very fact that the weather has become increasingly chaotic, that fireball sightings have increased so dramatically, and even frequent impacts are recorded around the globe, are clues that we are definitely moving into a cosmic dust stream as Clube describes, and that such a stream very likely includes some bigger objects – swarms of them – and those 'in the know' have taken note and are making preparations to survive. Shouldn't that knowledge be available to everyone? We think so. We don't even have to worry about anything having to do with a companion star and a comet swarm from the Oort cloud, which may or may not be in our future; we only have to deal with the science in front of us.

Based on what Clube writes in his letter, it seems that we are on our own, opposed by the governments that are supposed to be in place to look after the interests of their people. Of course, the question arises: what led to this general and overall blindness on the part of the people we look to for interpretation and explanation of our reality? How can the people who write textbooks, teach in schools, even at the highest level, be so ignorant? The consequences of this ignorance are, after all, detrimental to everyone for many reasons, not the least of which is simple survival in a rather hostile environment.

The events that have been covered so far in this book have led us to understand that there have been many times when it is highly probable that the earth – or parts thereof – was bombarded with meteorites or exploding aerial cometary fragments. These events occurred, and were probably related to, periods of great stress on the environment and humanity as a whole. Climate changes brought floods, droughts, extreme temperatures, crop failures and famine. These pressures may have caused lowered disease resistance for given populations, and it is also conjectured that extra-terrestrial bombardments may have carried disease pathogens. Impacts or crustal disturbances could have placed stresses on the geological structures so that outgassings from fissures, the ocean, or lakes may have poisoned large numbers of people, not to mention the record of tsunamis that is now called into question. Do we know, for example, that the Christmas tsunami-causing earthquake near Malaysia was not impact induced? No, we don't. And we can't trust either our governments or the news media – or even most of academia who owe their livelihoods to the government – to tell us the truth. Why do they lie to us?

Well, the main reason is rather simple: it's all about control. All of these things, taken together, place intolerable stresses on the human social organism and, as is typical for human beings, this brings on a crisis of faith, demands for answers, demands for protection that governments simply find too expensive to provide. When the world shows itself to be a hostile environment, when the environment suggests that there is no God and humanity is cast adrift in an uncaring cosmos, most people cannot tolerate this; they desperately need to restore their belief in something 'out there' that is going to save them, and if there is no one to fulfill this role, that means that someone has to be blamed for the disasters: a scapegoat. The corrupt governments do not want to be blamed, so they seek to blame someone else and convince the masses that this object of derision is the chief cause of all terrors. And the masses invariably buy into these maneuvers because, of course, if you can find someone or something to blame for calamity, you can continue in your illusion that 'God is in his heaven and – but for the evil acts of the chosen scapegoat – all would be right with the world'. I'm sure that you notice that this also relieves the individual of any responsibility, so this approach works in all kinds of situations.

We are going to examine this problem in some depth further on, but for now, I would like the reader to become acquainted with the facts. I have prepared a list, by no means exhaustive, of all the incidents I have been able to uncover of meteorite, asteroid, or cometary impacts that have caused death and destruction, property damage, or were near misses. Major parts of this list are extracted from the work of John S. Lewis, specifically his books *Rain of Iron and Ice* and *Comet and Asteroid Impact Hazards on a Populated Earth*.[25] As I already quoted in the introduction, in this latter volume, he writes:

> [T]he most intensively studied impact phenomenon, impact cratering, is of limited importance, due to the rarity and large mean time between events for crater-forming impacts. *Almost all events causing property damage and lethality are due to bodies less than 100 meters in diameter, almost all of which, except for the very largest and strongest, are fated to explode in the atmosphere. ...* [W]e are forced to conclude that the complex behavior of smaller bodies is closely relevant to *the threat actually experienced by contemporary civilization.* [emphasis added]

---

[25] The other major source for reports, especially from the last decade, is the *SOTT.net* archives. You can view the monthly digests of collected reports at *fireballs-meteorites. blogspot.ca/*

Based on the data he collected, Lewis noted that:

[O]n the century time scale, *firestorm ignition and direct blast damage by rare, strong, deeply penetrating bodies are the most common threats to human life*, with average fatality rates of about 250 people per year. ... On a 1000-year scale, the most severe single event, which is usually a 10 to 100-megaton Tunguska-type airburst, accounts for most of the total fatalities. On longer time scales, regional *impact-triggered tsunamis* become the most dangerous events. ... The exact impactor threshold size for global effects remains poorly determined. ... Perhaps most interesting is the implication that the large majority of lethal events (not of the number of fatalities) are *caused by bodies that are so small, so faint, and so numerous that the cost of the effort required to find, track, predict, and intercept them exceeds the cost of the damage incurred* by ignoring them. [emphasis added]

Unfortunately, Professor Lewis did not have at hand the information presented by Mike Baillie in his book *New Light on the Black Death*, nor did he consider the global events of 12,000 years ago revealed by the work of maverick scientists, Firestone, West and Warwick-Smith. If he had added the estimated numbers of fatalities from those events into his calculations, he might not have decided that the small, faint, and numerous bodies were so easily ignored. I think that if *all* the data were plugged in, the average deaths per year would be a lot higher than 250. Regarding impacts from history, Lewis writes in *Comet and Asteroid Impact Hazards on a Populated Earth*:

Many ancient sources from many cultures treat comets as literal, physical harbingers of doom. Such phenomena as the burning of cities and the overthrow of buildings and walls by aerial events are mentioned many times in Latin, Greek, Hebrew, and Chinese records, but there is no evidence of physical understanding of the nature of the bombarding objects or their effects until quite recently. ...

There is indeed a language problem in understanding the ancient reports, but it is largely a matter of the lack of an appropriate technical vocabulary in the older writings. ... *In certain locations and periods, especially in medieval Europe, all unusual heavenly events were interpreted as signs sent by God. Therefore, the surviving accounts are strongly biased toward explaining the moral purpose of these events, not their physical nature.* Such fundamental information as exact date and time, exact location, place of appearance of the phenomenon in the sky, its duration and physical extent, luminosity, precise nature of the damage done, and the like, were generally regarded as unimportant, and therefore rarely recorded for posterity. ... Even in 20th-century newspapers, bolide explosions may be described (and indexed) as "mysterious explosions," aerial blasts, aerolites, aeroliths, bolides, earthquakes, fireballs, meteorites, meteors, shocks, thunder, and so on. ...

Reports of meteorite falls, often with consequent damage, extend back to the fall of a "thunderstone" in Crete in 1478 BC, described by Malchus in the *Chronicle of Paros*. The earliest Biblical source is the account of a lethal fall of stones in … Joshua 10:11. … Other ancient reports in the West are found in the writings of Pausanius, Plutarch, Livy, Pindar, Valerius Maximus, Caesar, and many others. The report of a great fall of black dust at Constantinople in 472 BC, perhaps the result of a high-altitude airburst, is documented by Procopius, Ammianus Marcellinus, Theophanes, and others.

Colonel S. P. Worden has called to my attention the following passage in *The History of the Franks*, written by Bishop Gregory of Tours: "580 AD in Louraine, one morning before the dawning of the day, a great light was seen crossing the heavens, falling toward the east. A sound like that of a tree crashing down was heard over all the countryside, but it could surely not have been any tree, since it was heard more than fifty miles away … the city of Bordeaux was badly shaken by an earthquake … *a supernatural fire burned down villages* about Bordeaux. It took hold so rapidly that houses and even threshing-floors with all their grain were burned to ashes. Since there was absolutely *no other visible cause of the fire*, it must have happened by divine will. The city of Orleans also burned with so great a fire that even the rich lost almost everything."

Astronomers who have sought documentary evidence of ancient astronomical phenomena (eclipses, comets, fireballs, etc.) have found that East Asian records are far superior to European records for many centuries. Kevin Yau has searched Chinese records and found many reports of deaths and injuries (Yau *et al.*, 1994). The Chinese records of lethal impact events include the death of ten victims from a meteorite fall in 616 AD, an "iron rain" in the O-chia district in the 14th century that killed people and animals, several soldiers injured by the fall of a "large star" in Hot'ao in 1369, and many others. The most startling is a report of an event in early 1490 in Ch'ing-yang, Shansi, in which many people were killed when stones "fell like rain." Of the three known surviving reports of this event, one says that "over 10,000 people" were killed, and one says that "several tens of thousands" were killed.

On September 14, 1511, a meteorite fall in Cremona, Lombardy, Italy, reportedly killed a monk, several birds, and a sheep. In the 17th century we find reports of a monk in Milano, Italy, who was struck by a meteorite that severed his femoral artery, causing him to bleed to death, and of two sailors killed on shipboard by a meteorite fall in the Indian Ocean.

In addition to these shipboard fatalities, there have been several striking accounts of near disasters involving impacts very close to ships. Near midnight of February 24, 1885, at a latitude of 37 degrees N and a longitude of 170 degrees 15 minutes E in the North Pacific, the crew of the barque Innerwich, en route from Japan to Vancouver, saw the sky turn fiery red: "A large mass of fire appeared over the vessel, completely blinding the spectators; and, as it fell into the sea some 50 yards to leeward, it caused a hissing sound, which was heard above the blast, and made the vessel quiver from stem to stem. Hardly had this disappeared, when a lowering mass of white foam was seen rapidly approaching the vessel. The noise from the advancing volume of water is described as deafening. The barque was struck flat aback; but, before there was time to touch a brace, the sails had filled again, and the roaring white sea had passed ahead."

A strikingly similar event occurred only two years later on the opposite side of the world. Captain C. D. Swart of the Dutch barque J.P.A. reported in the *American Journal of Meteorology* 4 (1887) that, when sailing at 37 degrees 39 minutes N and 57 degrees W, at about 5pm on March 19, 1887, during a severe storm in which it was "as dark as night above," two brilliant fireballs appeared as in a sea of fire. One bolide "fell into the water very close alongside the vessel with a roar, and caused the sea to make tremendous breakers which swept over the vessel. *A suffocating atmosphere and perspiration ran down every person's face on board and caused everyone to gasp for fresh air. Immediately after this, solid lumps of ice fell on deck, and everything on deck and in the rigging became iced, notwithstanding that the thermometer registered 19 degrees Centigrade.*"

... Next, according to the *Times*, on September 13, 1930, a fireball plunged into the sea near Eureka, California, barely missing the tug Humboldt, which was towing the Norwegian motorship *Childar* out to sea. It requires little imagination to appreciate that such an event, if it were to strike a ship, should easily cause fatalities, or even the loss of the vessel with all hands. [emphasis added] (Lewis 1999, 1–5)

Now, that just gives you a taste of what is to come.

THE LIST:
Damages, Disasters, Injuries, Deaths, and Very Close Calls

■ 10000–11000 BCE (10700) – The earliest disaster we know of from our historical or mythic records is, of course, the legendary Deluge of Atlantis. The description of the end of Atlantis given by Plato in the Timaeus and Critias dialogues bears striking resemblance to what many scientists are now agreed would be the inevitable result of an oceanic impact by a disintegrating comet or large asteroid. The resultant tsunami, or tidal waves, would easily reach 2,000 feet high as they approached land, wiping out any and all coastal settlements. The deluge traditions, of which there are literally hundreds worldwide, appear in this light to be variations on Plato's account, and could even be actual observation-based tales, eyewitness accounts of the same, or similar, events. This is very likely the event discussed by Firestone, West and Warwick-Smith in *The Cycle of Cosmic Catastrophes: How a Stone-Age Comet Changed the Course of World Culture*. As I have discussed in my book, *The Secret History of the World*, the North and South American continents in the Western Hemisphere fit all the descriptions of 'Atlantis', and it is very likely that the event that led to the extinction of about 30 species of large mammals about 12,000 years ago was the source of the legends of Atlantis and probably the legends of a global deluge, such as Noah's Flood. Let's look at some descriptions of what such an event can do.

Back in the 1940s, Dr. Frank C. Hibben, Professor of Archeology at the University of New Mexico, led an expedition to Alaska to look for human remains. He didn't find human remains; he found miles and miles of icy muck just packed with mammoths, mastodons, and several kinds of bison, horses, wolves, bears and lions. Just north of Fairbanks, Alaska, the members of the expedition watched in horror as bulldozers pushed the half-melted muck into sluice boxes for the extraction of gold. Animal tusks and bones rolled up in front of the blades "like shavings before a giant plane". The carcasses were found in all attitudes of death, most of them "pulled apart by some unexplainable prehistoric catastrophic disturbance." (Hibben, Frank, *The Lost Americans*. New York: Thomas & Crowell Co. 1946)

The killing fields stretched for literally hundreds of miles in every direction. [*ibid*] There were trees and animals, layers of peat and moss, twisted and tangled and mangled together as though some Cosmic mixmaster sucked them all in circa 12,000 years ago, and then froze them instantly into a solid mass. (Sanderson, Ivan T., 'Riddle of the Frozen Giants', *Saturday Evening Post*, No. 39, January 16, 1960.)

Just north of Siberia entire islands are formed of the bones of Pleistocene animals swept northward from the continent into the freezing Arctic Ocean. One estimate suggests that some ten-million animals may be buried along the rivers of northern Siberia. Thousands upon thousands of tusks created a massive ivory trade for the master carvers of China, all from the frozen mammoths and mastodons of Siberia. The famous Beresovka mammoth first drew attention to the preserving properties of being quick-frozen when buttercups were found in its mouth.

What kind of terrible event overtook these millions of creatures in a single day? The evidence suggests an enormous tsunami raging across the land, tumbling animals and vegetation together, to be finally quick-frozen for the next 12,000 years. But the extinction was not limited to the Arctic, even if the freezing at colder locations preserved the evidence of Nature's rage.

Paleontologist George G. Simpson considers the extinction of the Pleistocene horse in North America to be one of the most mysterious episodes in zoological history, confessing "no one knows the answer." He is also honest enough to admit that there is the larger problem of the extinction of many other species in America at the same time. (Simpson, George G., *Horses*. New York: Oxford University Press, 1961) The horse, giant tortoises living in the Caribbean, the giant sloth, the saber-toothed tiger, the glyptodont and toxodon, these were all tropical animals. These creatures didn't die because of the "gradual onset" of an ice age; "unless one is willing to postulate freezing temperatures across the equator, such an explanation clearly begs the question." (Martin, P. S. & Guilday, J. E., 'Bestiary for Pleistocene Biologists', *Pleistocene Extinction*, Yale University, 1967)

Massive piles of mastodon and saber-toothed tiger bones were discovered in Florida. (Valentine, quoted by Berlitz, Charles, *The Mystery of Atlantis*, New York, 1969) Mastodons, toxodons, giant sloths and other animals were found in Venezuela quick-frozen in mountain glaciers. Woolly rhinoceros, giant armadillos, giant beavers, giant jaguars, ground sloths, antelopes and scores of other entire species were all totally wiped out at the same time, at the end of the Pleistocene, approximately 12,000 years ago.

This event was global. The mammoths of Siberia became extinct at the same time as the giant rhinoceros of Europe, the mastodons of Alaska, the bison of Siberia, the Asian elephants and the American camels. It is obvious that the cause of these extinctions must be common to both hemispheres, and that it was not gradual. A "uniformitarian glaciation" would not have caused extinctions because the various animals would have simply migrated to better pasture. What is seen is a surprising event of uncontrolled violence. (Leonard, R. Cedric, Appendix A in 'A Geological Study of the Mid-Atlantic Ridge', Special Paper No. 1, Bethany: Cowen Publishing, 1979) *In other words, 12,000 years ago, something terrible happened – so terrible that life on earth was nearly wiped out in a single day.*

Harold P. Lippman admits that the magnitude of fossils and tusks encased in the Siberian permafrost present an "insuperable difficulty" to the theory of uniformitarianism, since no gradual process can result in the preservation of tens of

thousands of tusks and whole individuals, "even if they died in winter." (Lippman, Harold E., "Frozen Mammoths", *Physical Geology*, New York 1969) Especially when many of these individuals have undigested grasses and leaves in their belly. Pleistocene geologist William R. Farrand of the Lamont-Doherty Geological Observatory, who is opposed to catastrophism in any form, states: "Sudden death is indicated by the robust condition of the animals and their full stomachs... the animals were robust and healthy when they died." (Farrand, William R., "Frozen Mammoths and Modern Geology", *Science*, Vol.133, No. 3455, March 17, 1961) Unfortunately, in spite of this admission, this poor guy seems to have been incapable of facing the reality of worldwide catastrophe represented by the millions of bones deposited all over this planet right at the end of the Pleistocene. Hibben sums up the situation in a single statement: *"The Pleistocene period ended in death. This was no ordinary extinction of a vague geological period, which fizzled to an uncertain end. This death was catastrophic and all inclusive."* (Hibben, *op. cit.*)
[Excerpted from Laura Knight-Jadczyk, *The Secret History of The World*]

Dr. Benny Peiser writes in the December 1997 edition of *British Archaeology*:

Until recently, the astronomical mainstream was highly critical of Clube and Napier's giant-comet hypothesis. However, the crash of comet Shoemaker-Levy 9 on Jupiter in 1994 has led to a change in attitudes. The comet, watched by the world's observatories, was seen split into 20 pieces and slammed into different parts of the planet over a period of several days. A similar impact on Earth, it hardly needs saying, would have been devastating.[26]

The Carolina Bays date to this time: mysterious land features often filled with bay trees and other wetland vegetation. Because of their oval shape and consistent orientation, they are considered by some authorities to be the result of a vast meteor shower that occurred approximately 12,000 years ago. What is most astonishing is the number of them. There are over 500,000 of these shallow basins dotting the coastal plain from Georgia to Delaware. That is a frightening figure.
Let me repeat: *There are over 500,000 of these shallow basins.*

Unlike virtually any other bodies of water or changes in elevation, these topographical features follow a reliable and unmistakable pattern. Carolina Bays are circular, typically stretched, elliptical depressions in the ground, oriented along their long axis from the Northwest to the Southeast. [T]hey are further characterized by an elevated rim of fine sand surrounding the perimeter. ...

---

[26] http://www.britarch.ac.uk/ba/ba30/Ba30feat.html

Robert Kobres, an independent researcher in Athens, Georgia, has studied Carolina Bays for nearly 20 years in conjunction with his larger interest in impact threats from space. His recent self-published investigations have profound consequences for Carolina Bay study and demand research by academia as serious, relevant and previously unexamined new information. The essence of Kobres' theory is that *the search for "debris," and the comparison of Bays with "traditional" impact craters, falsely and naively assumes that circular craters with extraterrestrial material in them are the only terrestrial evidence of past encounters with objects entering earth's atmosphere.*

Kobres goes a logical step further by assuming that forces associated with incoming bodies, principally intense heat, should also leave visible signatures on the earth. And, finally, that *physics does not demand that a "collision" of the bodies need necessarily occur to produce enormous change on Earth.* To verify that such encounters are possible outside of the physics lab, we need look no further than the so-called "Tunguska Event."

At the epicenter of the explosion lay not a large crater with a "rock" in it, as might be expected, but nothing more than a number of "neat oval bogs." The Tunguska literature generally mentions the bogs only in passing, since the researchers examining the site failed to locate any evidence of a meteorite and went on to examine other aspects of the explosion.

(*The Secret History of The World*)

Now, how many human deaths ought we to assign to this event? As Firestone *et al.* discuss, it was global in effect and the evidence of a sharply reduced population of not only animals, but humans, is there in the geological record. But what was the total human population? What kind of numbers can we plug into Lewis' calculations? Frankly, we don't know. Undoubtedly, multiplied millions of human beings perished at that time, along with the extinction of many animal species. One thing that seems certain is that if these numbers were included in Lewis' assessment, it would make a significant change in the 'average number of deaths per year'. Though, of course, this was a very big event, and those don't happen every year, or even every century. They happen on a scale of thousands of years and there hasn't been one like that for 12,000 years.

- **9100 BCE** – Extinction of the woolly mammoths.

- **7500 BCE** – Brings ice age to an end.

- **5900 BCE** – Metals found smelted 'naturally', which gives rise to humans smelting metals.

■ **4300 BCE** – Metals smelted naturally; beginning of Homeric 'religions'. Possible time for event on which part of the Exodus story is based.

■ **3195 BCE** – Eco-disaster as shown in tree rings. What evidence is there then that something unusual happened around 3100 BCE, other than the Mayan year-zero, supposedly relating to 3114 BCE? We have the construction of Newgrange in Ireland; floods in the paleoclimatic data; the construction of Stonehenge number one; the unification of Egypt; methane peak fires; cold weather, according to bristlecone pines; and the erection of the coastal menhirs in Brittany. Although any one of these in itself would not be unusual, the timing of them within a frame of only 100 years is what makes us suspect that something unusual was going on. The next 1,000 years or so were a very restless time globally:

> The postulated bombardments and dust-veils at around 3195 BCE, another narrowest tree-ring date, would have wreaked havoc on both the local and global climate, and any and all cultures affected would have taken many decades, maybe even centuries, to recover. The sheer terror that "multiple-Tunguska-class fireballs" would have instilled into the peoples of those times would have understandably motivated them towards building some form of observatories to help predict future meteor showers/storms as a matter of perceived urgency …
>
> (*The Secret History of The World*)

Stonehenge may very well have been built to help in the watch for comets (an idea supported and developed by Christopher Knight and Robert Lomas, in their book, *Uriel's Machine*). And, yet again, we have no numbers of human fatalities to plug into the calculations, but they must have been enormous.

■ **3123 BCE, June 29** – Germany: 'The clay tablet that tells how an asteroid destroyed Sodom 5,000 years ago': [27]

> A clay tablet that has baffled scientists for more than a century has been identified as a witness's account of an asteroid that destroyed the Biblical cities of Sodom and Gomorrah 5,000 years ago. Researchers believe that the tablet's symbols give a detailed account of how a mile-long asteroid hit the region, causing thousands of deaths and devastating more than one million sq km (386,000 sq miles). The impact, equivalent to more than 1,000 tons of TNT exploding, would have created one of the world's biggest-ever landslides. …

---

[27] http://www.dailymail.co.uk/pages/live/articles/news/news.html?in_article_id=551010 &in_page_id=1770

The theory is the work of two rocket scientists – Alan Bond and Mark Hempsell – who have spent the past eight years piecing together the archaeological puzzle. At its heart is a clay tablet called the Planisphere, discovered by the Victorian archaeologist Henry Layard in the remains of the library of the Royal Palace at Nineveh. Using computers to recreate the night sky thousands of years ago, they have pinpointed the sighting described on the tablet – a 700 BC copy of notes of the night sky as seen by a Sumerian astrologer in one of the world's earliest-known civilisations – to shortly before dawn on June 29th in the year 3123 BC. Half the tablet records planet positions and clouds, while the other half describes the movement of an object looking like a "stone bowl" travelling quickly across the sky. The description matches a type of asteroid known as an Aten type, which orbits the Sun close to the Earth. Its trajectory would have put it on a collision course with the Otz Valley. [In Germany, in other words. In short, the story wasn't about Abraham and Lot in Palestine!]

"It came in at a very low angle – around six degrees – and then clipped a mountain called Gaskogel around 11 km from Köfels," said Mr. Hempsell. "This caused it to explode – and as it travelled down the valley it became a fireball. When it hit Köfels it created enormous pressures which pulverised the rock and caused the landslide. But because it wasn't solid, *there was no crater.*" The explosion would have created a mushroom cloud, while a plume of smoke would have been seen for hundreds of miles.

Mr. Hempsell said another part of the tablet, which is 18 cm across and shaped like a bowl, describes a plume of smoke around dawn the following morning. "You need to know the context before you can translate it," said Mr. Hempsell, of Bristol University.

Geologists have dated the landslide to around 9,000 years ago, far earlier than the Sumerian record. However, Mr. Hempsell, who has published a book on the theory, believes *contaminated samples from the asteroid may have confused previous dating attempts.* Academics were also quick to disagree with the findings, which were published in *A Sumerian Observation of the Köfels's Impact Event.* John Taylor, a retired expert in Near Eastern archaeology at the British Museum, said there was no evidence that the ancient Sumerians were able to make such accurate astronomical records, while our knowledge of Sumerian language was incomplete. "I remain unconvinced by these results," he added. [emphasis added]

■ **2345 BCE** – Eco-disaster focused in the Levant as shown in tree-rings. End of Egyptian Old Dynasty?

The French archaeologist, Marie-Agnes Courty, presented a paper at the Society for Inter-Disciplinary Studies July 1997 conference at Cambridge University, in which she first detailed the findings of excavations at a site in northern Syria, at

Tell Leilan. This was the first time ever that an archaeological excavation had been initiated where the main purpose was to examine the stratigraphical record of the area with a view to searching for evidence of "scorched earth" due to a suspected episode of extra-terrestrial "fireball bombardment."

She and her team found much evidence of microscopic glass spherules typical of melted sand and rock which is caused by the intense heat resulting from an asteroid impact or air-burst. She recommended further excavations there and at other sites. It would make sense that attention should be focussed on sites once occupied at dates where the tree-ring chronologies show evidence of abrupt climate changes – as at Tell Leilan in northern Syria, where the "burn event" has now been dated by Courty as immediately prior to 2345 BC, a "narrowest tree-ring" date.

Another with no human fatality numbers included in the calculations.

Scientists have found the first evidence that a devastating meteor impact in the Middle East might have triggered the mysterious collapse of civilisations more than 4,000 years ago.

Studies of satellite images of southern Iraq have revealed a two-mile-wide circular depression which scientists say bears all the hallmarks of an impact crater. If confirmed, it would point to the Middle East being struck by a meteor with the violence equivalent to hundreds of nuclear bombs. Today's crater lies on what would have been shallow sea 4,000 years ago, and any impact would have caused devastating fires and flooding. The catastrophic effect of these could explain the mystery of why so many early cultures went into sudden decline around 2300 BC. The crater's faint outline was found by Dr Sharad Master, a geologist at the University of Witwatersrand, Johannesburg, on satellite images of the Al 'Amarah region, about ten miles north-west of the confluence of the Tigris and Euphrates and home of the Marsh Arabs. (Robert Matthews, Science Correspondent, *The Telegraph*, 11-4-1)

■ **1628 BCE** – 'The Exodus': Biblical scholars have been debating the date of the so-called Exodus for hundreds of years. The most recent researchers have indicated that there was no 'exodus' as depicted in the Bible, it was all made up by post-exilic priests to create a 'history' justifying their elite status and privileges. More than that, based on historical knowledge of how things were done in those times, they probably were not even related to any of the people 'carried away to Babylon' in the first place. And so, it seems logical to speculate that the background information contained in the Exodus story – and other related stories in the Bible, such as the collapse of Jericho and the destruction of Sodom and Gomorrah – were legendary stories of events that occurred around the time of the eruption of Thera, which has been fairly securely fixed around 1600 BCE, plus or minus fifty years. Mike Baillie reports that whatever happened during this period of history that includes this monstrous eruption, it was global in effect as is shown in the tree-ring chronologies. In other words, more was going on than just a volcanic eruption. Again, no numbers of fatalities to plug into the calculations, though there are many ancient reports of plague and mass death, and Egyptian records report many strange sky, weather and plague phenomena.

■ 1159 BCE – Collapse of Shang and Mycenean cultures. Collapse of the Bronze Age in the Mediterranean region. Possible origin of more 'Exodus' stories amalgamated with older tales of similar events. Wikipedia tells us:

> The Bronze Age collapse is the name given by those historians who see the transition from the Late Bronze Age to the Early Iron Age, as violent, sudden and culturally disruptive, expressed by the collapse of palace economies of the Aegean and Anatolia, replaced after a hiatus by the isolated village cultures of the Dark Age period of history of the Ancient Middle East.

A medieval painting depicting a 'rain of fire'.

Mike Baillie points out that a series of impacts/overhead explosions would more adequately explain the longstanding problem of the end of the Bronze Age in the Eastern Mediterranean in the 12th century BCE. At that time, many – uncountable – major sites were destroyed and totally burned, and it has all been blamed on those supernatural 'Sea Peoples'. If that was the case, if it was invasion and conquest, there ought to at least be some evidence for that, like dead warriors or signs of warfare; but for the most part, that is not the case. There were almost no bodies found, and no precious objects except those that were hidden away as though someone expected to return for them, or didn't have time to retrieve them. The people who fled (extra-terrestrial events often have precursor activities and warnings because a comet can often be observed approaching for some time) were probably also killed in the act of fleeing, and the result was total abandonment and total destruction of the cities in question. (John Lewis did not include this in his calculations either.)

■ **207 BCE** – 'Scientists Say Comet Smashed into Southern Germany in 200 BCE':[28]

> A comet or asteroid smashed into modern-day Germany some 2,200 years ago, unleashing energy equivalent to thousands of atomic bombs, scientists reported on Friday. The 1.1-kilometre (0.7-mile) diameter rock whacked into southeastern Bavaria, leaving an "exceptional field" of meteorites and impact craters that stretch from the town of Altoetting to an area around Lake Chiemsee, the scientists said in an article in the latest issue of US magazine Astronomy.
>
> Colliding with the Earth's atmosphere at more than 43,000 km per hour, the space rock probably broke up at an altitude of 70 km, they believe. The biggest chunk smashed into the ground with a force equivalent to 106 million tonnes of TNT, or 8,500 Hiroshima bombs. "The forest beneath the blast would have ignited suddenly, burning until the impact's blast wave shut down the conflagration," the investigators said. "Dust may have been blown into the stratosphere, where it would have been transported around the globe easily…. The region must have been devastated for decades."
>
> The biggest crater is now a circular lake called Tuettensee, measuring 370 metres (1,200 feet) across. Scores of smaller craters and other meteorite impacts can be spotted in an elliptical field, inflicted by other debris.
>
> The study was carried out by the Chiemgau Impact Research Team, whose five members included a mineralogist, a geologist and an astronomer. …
>
> Additional evidence comes from local discoveries of Celtic artifacts, which appear to have been scorched on one side. That helped to establish an approximate

---

[28] http://www.spacedaily.com/news/comet-04l.html

date for the impact of between 480 and 30 BCE. The figure may be fine-tuned to around 200 BCE, thanks to tree-ring evidence from preserved Irish oaks, which show a slowing in growth around 207 BCE. This may have been caused by a veil of dust kicked up by the impact, which filtered out sunlight. In addition, Roman authors at about the same time wrote about showers of stones falling from the skies and terrifying the populace.

The object is more likely to have been a comet than an asteroid, given the length of the ellipse and scattered debris, the report says.

■ **44 BCE** – Pliny states that *"Portentous and protracted eclipses of the sun occurred, such as the one after the murder of Caesar the dictator."* Yet there were no solar eclipses visible from anywhere in the Roman Empire from February of 48 BCE through December of 41 BCE, inclusive. There was a spectacular daylight comet in 44 BCE, perhaps the most famous comet in antiquity. A dust veil occluded the sky over Italy in the spring of 44, and has often been attributed to an (unconfirmed) eruption of Mt Etna. There are sulfate deposits in the Greenland ice cores for this year and there is tree ring evidence from North America, where dendrochronology points to a climatic change in the late 40s BCE. What hit and where it hit, has yet to be determined, and whether or not there was death and destruction somewhere on the globe, is unknown. (John S. Lewis does not include this event in his calculations.)

■ **60–70 AD** – The destruction of Jerusalem.

> The story Josephus tells of the sixties is one of famine, social unrest, institutional deterioration, bitter internal conflicts, class warfare, banditry, insurrections, intrigues, betrayals, bloodshed, and the scattering of Judeans throughout Palestine. ... There were wars and rumors of wars for the better part of ten years and *Josephus reports portents, including a brilliant daylight in the middle of the night!* (Burton Mack, *A Myth of Innocence: Mark and Christian Origins*, 1988) [emphasis added]

We recognize that 'brilliant daylight at night' from the Tunguska Event. Josephus gives several portents of the evil to befall Jerusalem and the temple. He described a star resembling a sword, a comet, a light shining in the temple, a cow giving birth to a lamb at the moment it was to be sacrificed in the Jerusalem Temple, *armies fighting in the sky*, and a voice from the Holy of Holies declaring, *"We are departing"* (Josephus, *Jewish Wars*, 6). (Obviously, the voice was apocryphal.) Some of these portents are mentioned by other contemporary historians, Tacitus for example. However, in book five of his *Histories*, Tacitus castigated the superstitious Jews for not recognizing and offering expiations for the portents to avert the disasters. He put the destruction of Jerusalem down to the stupidity or willful ignorance of the Jews themselves in not offering the appropriate sacrifices. *"Thus there was a star resembling a sword, which stood over the city [Jerusalem], and a comet, that continued a whole year"* (Josephus, *Jewish Wars*, 6.3).

In short, it very well may be that the eschatological writings in the New Testament – the very formation of the Myth of Jesus – was based on cometary events of the time, including a memory of the 'Star in the East'. The destruction of the Temple at Jerusalem may very well have been an 'Act of God', as reported by Mark in his Gospel.

■ **312 AD** – Italy: A team of geologists believes it has found the incoming space rock's impact crater, and dating suggests its formation coincided with the celestial vision said to have converted a future Roman emperor to Christianity. The small circular "Cratere del Sirente" in central Italy is clearly an impact crater, said the geologists, because its shape fits and it is also surrounded by numerous smaller, secondary craters, gouged out by ejected debris, as expected from impact models.

Radiocarbon dating puts the crater's formation at about the right time to have been witnessed by Constantine and there are magnetic anomalies detected around the secondary craters – possibly due to magnetic fragments from the meteorite. It would have struck the earth with the force of a small nuclear bomb, perhaps a kiloton in yield. It would have looked like a nuclear blast, with a mushroom cloud and shockwaves.

■ **476 AD** – I-hsi and Chin-ling, China: 'Thundering chariots' fell to ground 'like granite'; vegetation was scorched.

■ **526 AD** – Great Antioch earthquake.

> … those caught in the earth beneath the buildings were incinerated, and sparks of fire appeared out of the air and burned everyone they struck like lightning. The surface of the earth boiled and foundations of buildings were struck by thunderbolts thrown up by the earthquakes, and were burned to ashes by fire. … It was a tremendous and incredible marvel with fire belching out rain, rain falling from tremendous furnaces, flames dissolving into showers … as a result Antioch became desolate … in this terror up to 250,000 people perished. (Malalas, quoted in Baillie 2006, 130)

■ **536–45 AD** – Reduced sunlight, mists or 'dry fogs', crop failures, plagues, and famines in China and the Mediterranean. The Roman Praetorian Prefect, Magnus Aurelius Cassiodorus, wrote a letter documenting the conditions:

All of us are observing, as it were, a blue-coloured sun; we marvel at bodies which cast no mid-day shadow, and at that strength of intensest heat reaching extreme and dull tepidity... So we have had a winter without storms, spring without mildness, summer without heat... The seasons have changed by failing to change; and what used to be achieved by mingled rains cannot be gained from dryness only.

Procopius of Caesarea, a Byzantine, wrote:

And it came about during this year that a most dread portent took place. For the sun gave forth its light without brightness, like the moon, during the whole year, and it seemed exceedingly like the sun in eclipse, for the beams it shed were not clear nor such as it is accustomed to shed.

John of Ephesus, cleric and a historian, wrote:

The sun was dark and its darkness lasted for eighteen months; each day it shone for about four hours; and still this light was only a feeble shadow... the fruits did not ripen and the wine tasted like sour grapes.

In the wake of this inexplicable darkness, crops failed and famine struck. And then, pestilence. But here we mean 'pestilence' as Jacme d'Agramaont, a doctor writing in 1348, described it in reference to the 'Black Death': *"the time of tempest caused by light from the stars."* During the time of Justinian, this pestilence ravaged Europe, *reducing the population of the Roman Empire by a third*, killing four-fifths of the citizens of Constantinople, reaching as far East as China and as far northwest as Great Britain. John of Ephesus documented the progress of this pestilence in 541-542 AD in Constantinople, where city officials gave up trying to count the dead after two-hundred and thirty thousand:

The city stank with corpses as there were neither litters nor diggers, and corpses were heaped up in the streets. ... It might happen that [a person] went out to market to buy necessities and while he was standing and talking or counting his change, suddenly the end would overcome the buyer here and the seller there, the merchandise remaining in the middle with the payment for it, without there being either buyer or seller to pick it up.

This was also the time assigned to the legendary King Arthur, the loss of the Grail, and the manifestation of the Wasteland. Although scholars place the historical King Arthur in the 5th century, the date of his death is given as 539 AD. According to Mike Baillie, *the imagery from the Arthurian legend is in accordance with the appearance of a comet and subsequent famine and plague*:

the 'wasteland' of legend. Ireland's St. Patrick stories feature a wasteland as well. And although St. Patrick is credited with ridding Ireland of snakes, we might consider that there never were snakes in Ireland, and that snakes and dragons are images associated with comets.

Until that point in time, the Britons had held control of post-Roman Britain, keeping the Anglo-Saxons isolated and suppressed. After the Romans

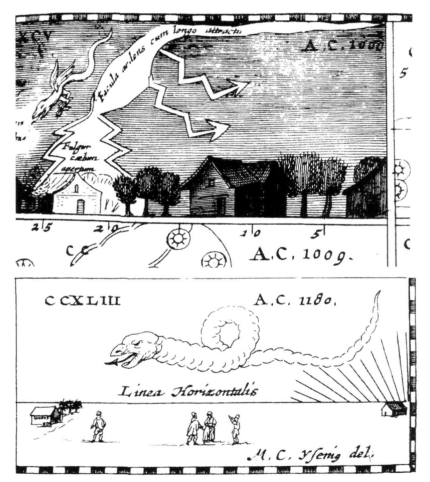

Illustrations of two important historical comets taken from Lubienietski's Universal History of All Comets, *1681. The AD 1000 illustration shows a blazing thunderbolt with a 'long drawn out tail landing in open space', having 'fallen from a dragon-like comet with a horrendous tail'. The AD 1180 comet was viewed with horror since it had the appearance of a winding serpent with gaping jaws.*

were gone, the Britons maintained the *status quo*, living in towns, with elected officials, and carrying on trade with the empire. After 536 CE, the year reported as the 'death of Arthur', the Britons, the ancient Cymric empire that at one time had stretched from Cornwall in the south to Strathclyde in the north, all but disappeared, and were replaced by Anglo-Saxons. There is much debate among scholars as to whether the Anglo-Saxons killed all of the Britons, or assimilated them. Here we must consider that they were victims of possibly many overhead cometary explosions which wiped out most of the population of Europe, plunging it into the Dark Ages which were, apparently, literally *dark*, atmospherically speaking.

The mystery of the origins of the red dragon symbol, now on the flag of Wales, has perplexed many historians, writers, and romanticists, and the archæological community generally has refrained from commenting on this most unusual emblem, claiming it does not concern them. In the ancient Welsh language it is known as *Draig Goch*, "red dragon"; and in ... the *University of Wales Welsh Dictionary* (University of Wales Press, 1967, p. 1082) there are translations for the various uses of the Welsh word draig. Amongst them are common uses of the word, which is today taken just to mean a "dragon," but in times past it has also been used to refer to *Mellt Distaw* (sheet lightning) and also *Mellt Didaranau* (lightning unaccompanied by thunder).

But the most interesting common usage of the word in earlier times, according to this authoritative dictionary, is *Maen Mellt*, the word used to refer to a "meteorite." And this makes sense, as the Welsh word *maen* translates as "stone," while the Welsh word *mellt* translates as "lightning" – so literally a "lightning-stone." That the ancient language of the Welsh druids has words still in use today which have in the past been used to describe both a dragon and also a meteorite, is something that greatly helps us to follow the destructive "trail of the dragon" as it was described in early Welsh "riddle-poems." ...

The exact nature and sequence of events in mid 6th-century AD that gave rise to the period we refer to as the European "Dark Age" is still a matter for speculation amongst historians and archaeologists. Over the past twenty years or so, certain paleo-climatologists have begun comparing notes with archaeologists and astronomers, and interestingly, in the absence of written records, many have begun to look a little more closely at mythology in their efforts to corroborate the findings of their researches. While much of this recent bout of inter-disciplinary brainstorming has focussed on the 6th century AD start of the European Dark Age, earlier dates are also of great interest to those embroiled in this veritable "paradigm shift". ...

In recent years certain astronomers have increasingly come to appreciate that encoded in the folklore and mythologies of many cultures are the accurate observations of ancient skywatchers. Almost all tell of times when death and mass destruction came from the skies, events that are often portrayed as "celestial battles" between what they variously depicted as "the Gods." And curiously the imagery in these "myths" have many common features, even between the mythologies of cultures widely spaced in time and location.[29]

Out on the Asian steppes, whatever happened in 536 CE caused political upheaval. The horse-based economy of the war-like Avars foundered, and their vassals, the cattle-herding Turks, overthrew them. Driven from the steppes, the Avars joined forces with the Slavs in Hungary on the borders of the Roman Empire.

Gildas, who was writing at approximately 540 CE, says that *"the island of Britain **was on fire from sea to sea** ... until it had burned almost the whole surface of the island and was licking the western ocean with its fierce red tongue."* In *The Life of St. Teilo*, contained in the Llandaf Charters of St. Teilo, who had recently been made Bishop of Llandaf Cathedral in Morganwg, South Wales, it says:

> [H]owever he could not long remain, on account of the pestilence which nearly destroyed the whole nation. It was called the Yellow Pestilence, because it occasioned all persons who were seized by it to be yellow and without blood, and it appeared to men a column of a watery cloud, having one end trailing along the ground, and the other above, proceeding in the air, and passing through the whole country like a shower going through the bottom of valleys. Whatever living creatures it touched with its pestiferous blast, either immediately died, or sickened for death ... and so greatly did the aforesaid destruction rage throughout the nation, that it caused the country to be nearly deserted.

St. Teilo is recorded as having left South Wales for Brittany to escape the Yellow Pestilence, and that it lasted for some eleven years.

In 540, in Yemen, the Great Dam of Marib, dating from around the 7th century bce, one of the engineering wonders of the ancient world and a central part of the south Arabian civilization, broke and began to collapse. By 550, the dam was a complete loss and thousands of people migrated to another oasis on the Arabian peninsula, Medina. The Arab tribes, traumatized by the environmental disasters around them, began to think of conquest for the sake of survival. In 610, a new leader unified them: Muhammad.

---

[29] 'The European "Dark Age" and Welsh Oral Tradition': http://www.morien-institute.org/darkages.html

Although a great many historical changes happened in the seventh century, such as the Roman war with Persia, the rise of Islam, rebellion and civil war in the Roman empire, and the advance of the Slavs driven by the Avars, it can be said that the seeds of these changes, the destruction of the old that made way for the new, can be traced to the environmental catastrophe of 536.

John Lewis does not include any estimates of the death and destruction occurring at that time in his 'average number of annual deaths by comets'.

■ **580 AD** – France: Great fireball and blast; Orleans and nearby towns burned.

■ **588 AD** – China: 'Red-colored object' fell with 'noise like thunder' into furnace, exploded, burned several houses.

■ **616 AD** – China: Ten deaths reported in China from a meteorite shower; seige towers destroyed.

■ **679 AD** – Coldingham, England: Monastery destroyed by 'fire from heaven' as reported in the *Anglo-Saxon Chronicle*.

■ **764 AD** – Nara, Japan: Meteorite strikes house.

■ **810 AD** – Upper Saxony: Charlemagne's horse startled by meteor, throws him to the ground.

■ **1000 AD** – Alberta, Canada: The date of this meteor strike is estimated:

> What local hunters in Whitecourt thought for years was a sinkhole is actually the crater left behind by a meteor that fell to earth 1,000 years ago and is now attracting international attention from researchers. ... The crater is 36-metres wide and six-metres deep, which is small as far as most craters go. ... Herd thinks the meteor came from the asteroid belt and measured one-metre across. However, researchers have so far found 74 different pieces of the original meteor – which is called a meteorite once it hits the ground – scattered around the crater, some up to 70 metres away.[30]

■ **1064** – Chang-chou, China: Daytime fireball, meteorite fall; fences burned.

■ **1178 (June)** – Canterbury, England:

---

[30] http://www.sott.net/articles/show/169469

In this year, on the Sunday before the Feast of St. John the Baptist, after sunset when the moon had first become visible a marvelous phenomenon was witnessed by some five or more men who were sitting there facing the moon. Now there was a bright new moon, and as usual in that phase its horns were tilted toward the east; and suddenly the upper horn split in two. From the midpoint of the division a flaming torch sprang up, spewing out, over a considerable distance, fire, hot coals, and sparks. Meanwhile the body of the moon which was below writhed, as it were in anxiety, and to put it in the words of those who reported it to me and saw it with their own eyes, the moon throbbed like a wounded snake. Afterwards it resumed its proper state. This phenomenon was repeated a dozen times or more, the flame assuming various twisting shapes at random and then returning to normal. Then after these transformations the moon from horn to horn, that is along its whole length, took on a blackish appearance. The present writer was given this report by men who saw it with their own eyes, and are prepared to stake their honour on an oath that they have made no addition or falsification in the above narrative. (Gervase of Canterbury)

■ **1321–68** – O-chia District, China: 'Iron rain' kills people, animals, damages house.

■ **1347–48** – Europe: The Black Death – not included in John Lewis' calculations – killed about half the population of Western Europe. The effects of this event were possibly global, though the number of deaths worldwide is unknown.

■ **1348 (January)** – Carinthia, Austria: Earthquake, 16 cities destroyed, fire fell from heaven; over 40,000 dead. John Lewis does not include this event in his calculations.

■ **1369** – Ho-t'ao, China: 'Large star' fell, starts fire, soldiers injured.

■ **1430 (June)** – New Zealand: A huge comet struck the ocean less than a hundred miles from the Chinese fleet of Zhou Man. Up to 173 wrecks have been counted as destroyed by this event. The comet incinerated many ships and hurled the blazing wrecks onto New Zealand South Island and the east coasts of Australia, and across the Pacific and Indian Oceans. Chinese and Mayan astronomers describe a large blue comet seen in *Canis Minor* for 26 days in June 1430 – a date compatible with Professor Bryant's (1410–1480) evidence. In November 2003 Dallas Abbott and her team announced they had found where the comet crashed – between Campbell Island and New Zealand South Island. Deaths? Probably in the multiple thousands:

A mega-tsunami struck southeast Asia 700 years ago rivaling the deadly one in 2004, two teams of geologists said after finding sedimentary evidence in coastal marshes. Researchers in Thailand and Indonesia wrote in two articles in *Nature* magazine that the tsunami hit around 1400, long before historical records of earthquakes in the region began.[31]

Which of course, leads us to ask the question: Was the 2004 mega-tsunami the result of a comet/asteroid strike?

■ **1490 (February)** – Ch'ing-yang, Shansi, China: Stones fell like rain; more than 10,000 killed.

■ **1492** – Ensisheim, Alsace: 280-pound meteorite landed; in the same year Columbus reported 'a marvelous branch of fire' that fell into the sea as he crossed the Atlantic.

■ **1511 (September 5)** – Cremona, Lombardy, Italy: Monk killed with several birds, a sheep.

■ **1516 (May)** – Nantan, China: *"During summertime in May of Jiajing 11th year, stars fell from the northwest direction, five to six fold long, waving like snakes and dragons. They were as bright as lightning and disappeared in seconds."* Many of them were recovered by local farmers in 1958 when China needed steel for the 'Great Leap Forward' advocated by Mao Zedong. They have coarse octahedral structure and contain 92.35% iron and 6.96% nickel, belonging to IIICD classification of Wasson *et al* (1980). Most Nantan meteorites weigh 150 to 1500 kg. Due to the humid conditions, smaller pieces buried in soils of lower valleys have been extensively weathered and oxidized into limonite.

■ **1620** – Punjab, India: 'Hot iron' fell, burned grass; made into dagger knife, two sabres.

■ **1631** – Fall of Magedeburg, Germany:

[A] grand storm-wind picked up, the town was inflamed at all possible places, so that even little aid (rescue) was of help (appreciated). ... Then I saw the whole town of Magdeburg, except dome, monastery and New Market, lying in embers

---

[31] http://www.sott.net/articles/show/168328

and ashes, which raged only about 3 or 3-1/2 hours, from which I deduced God's strange omnipotence and punishment. (Geoffrey Mortimer, 'Style and Fictionalisation in Eyewitness Personal Accounts of the Thirty Years War', *German Life and Letters*, 54:2)

A 'second sun' was seen on and around 29 May 1630; and on 20 May 1631, one year later, Magdeburg fell as described above. The standard historical description of the Fall of Magdeburg goes pretty much as follows:

The fall of Magdeburg horrified Europe. The city had been starved and then was bombarded unmercifully. The artillery shelling grew so bad, the town caught on fire. Over 20,000 of the citizens perished in the siege and the cataclysm that ended it. The city itself was burned to the ground. The cruel and pointless devastation marked a new low, an act abhorred by a generation well accustomed to horrors.

■ **1639** – China: Large stone fell in market; tens killed; tens of houses destroyed.

■ **1648** – Malacca: Two sailors reported killed on board ship en route from Japan to Sicily.

■ **1654** – Milano, Italy: Monk reported killed by meteorite.

■ **1661 (August 9)** – China: Meteorite smashes through roof; no injuries.

■ **1670 (November 7)** – China: Meteorite falls, breaks roof beam of house.

■ **1761** – Chamblan, France: House struck and burned by meteorite.

■ **1780** – New England, U.S./Eastern Ontario, Canada:

In the midst of the Revolutionary War, darkness descends on New England at midday. Many people think Judgment Day is at hand. It will be remembered as New England's Dark Day.

Diaries of the preceding days mention smoky air and a red sun at morning and evening. Around noon this day, an early darkness fell. Birds sang their evening songs, farm animals returned to their roosts and barns, and humans were bewildered.

Some went to church, many sought the solace of the tavern, and more than a few nearer the edges of the darkened area commented on the strange beauty of the preternatural half-light. One person noted that clean silver had the color of brass.

It was darkest in northeastern Massachusetts, southern New Hampshire and southwestern Maine, but it got dusky through most of New England and as far away as New York. At Morristown, New Jersey, Gen. George Washington noted it in his diary.

In the darkest area, people had to take their midday meals by candlelight. A Massachusetts resident noted, "In some places, the darkness was so great that persons could not see to read common print in the open air." In New Hampshire, wrote one person, "A sheet of white paper held within a few inches of the eyes was equally invisible with the blackest velvet."

At Hartford, Col. Abraham Davenport opposed adjourning the Connecticut legislature, thus: "The day of judgment is either approaching, or it is not. If it is not, there is no cause of an adjournment; if it is, I choose to be found doing my duty."

When it was time for night to fall, the full moon failed to bring light. Even areas that had seen a pale sun in the day could see no moon at all. No moon, no stars: It was the darkest night anyone had seen. Some people could not sleep and waited through the long hours to see if the sun would ever rise again. They witnessed its return the morning of May 20. Many observed the anniversary a year later as a day of fasting and prayer.

Professor Samuel Williams of Harvard gathered reports from throughout the affected areas to seek an explanation. A town farther north had reported "a black scum like ashes" on rainwater collected in tubs. A Boston observer noted the air smelled like a "malt-house or coal-kiln." Williams noted that rain in Cambridge fell "thick and dark and sooty" and tasted and smelled like the "black ash of burnt leaves."

As if from a forest fire to the north? Without railroad or telegraph, people would not know: No news could come sooner than delivered on horseback, assuming the wildfire was even near any European settlements in the vast wilderness.

But we know today that the darkness had moved southwest at about 25 mph. And we know that forest fires in Canada in 1881, 1950 and 2002 each cast a pall of smoke over the northeastern United States.

A definitive answer came in 2007. In the *International Journal of Wildland Fire*, Erin R. McMurry of the University of Missouri forestry department and co-authors combined written accounts with fire-scar evidence from Algonquin Provincial Park in eastern Ontario to document a massive wildfire in the spring of 1780 as the "likely source of the infamous Dark Day of 1780." [32]

Sounds like an impact event. If so, it would match the fourth period of cometary activity referenced by Clube.

---

[32] http://www.wired.com/science/discoveries/news/2008/05/dayintech_0519

■ **1790 (July 24)** – Barbotan and Agen, Gascony, France: Meteorite crushes cottage, kills farmer and some cattle.

■ **1794 (June 16)** – Siena, Italy: Child's hat hit; child uninjured.

■ **1798 (December 19)** – Benares, India: Building struck by meteorite.

■ **1801 (October 30)** – Suffolk, England: *"Dwelling-house of Mr. Woodrosse, miller, near Horringermill, Suffolk, was set on fire by a meteor, and entirely consumed, together with a stable adjoining."*

■ **1803 (July)** – East Norton, England: White Bull public house struck, chimney knocked down, grass burned, flight of object nearly horizontal.

■ **1803 (December 13)** – Massing, Czech Republic: Building struck by meteorite.

■ **1810 (July)** – Shahabad, India: Great stone fell, five villages burned, several killed.

■ **1811–12** – Eastern North America: Submitted by a reader, from *The Comet Book: A Guide For the Return of Halley's Comet* by Robert D. Chapman and John C Brandt, published in 1984:

> In December, 1811, a series of earthquakes began that rocked over three hundred thousand square miles of eastern North America. The great New Madrid Earthquake was just one of the many events of late 1811 and 1812 that followed quickly on the heels of the Great Comet of 1811. First observed from America in late 1811, the comet was described as bright and slightly smaller than the full moon, including its tail. Newspapers in the young republic picked up on the comet and predicted that it was an omen of evil times. Sure enough, a string of disasters, nature and otherwise, followed.
>
> Though the epicenter of the earthquakes was near a small town on the Mississippi River – New Madrid, Missouri – the numerous shocks were felt as far away as New York and Florida. According to one source, Richmond, Virginia and Boston were shaken so violently by the December 16 shocks that church bells rang. It is said that in some places the current in the Mississippi River flowed backward.
>
> The great naturalist John James Audubon was in Kentucky when the first tremor struck. Initially he thought the roar was a tornado and headed for shelter. Audubon and others reported strange darkenings and brightenings in the sky.

Jared Brookes, a native of Louisville, Kentucky kept accurate records of all the shocks that he experienced. Between December 1811 and May 1812, he tabulated over two thousand separate tremors. Based on the historical evidence, Charles Richter (inventor of the Richter scale used to measure the strength of earthquakes) had estimated that there were at least three severe shocks that exceeded magnitude eight on his scale. The New Madrid Earthquakes thus stand as one of the most severe, if not the most severe, series of quakes recorded in U.S. history.

The New Madrid Earthquakes were not the only disaster to take place in 1811 and 1812. On December 26, 1811, a fire broke out in the new theater in Richmond, which was packed with people. Governor George Smith and almost eighty others perished. The incidents leading to the War of 1812 were moving inexorably forward: the Battle of Tippecanoe; the Guerriere incident in which the British impressed an American seaman from an American vessel onto their warship Guerriere. To round out the problems, severe weather plagued the young republic. All told, 1812 was not a good year.

An interesting contrast to the disasters of 1811–1812 comes from the world of wine. The year 1811 produced a particularly good vintage. In honor of the great comet, the wine was referred to as *vin de la comète* (comet wine). Not everything from the year was bad!

Robert Fritzius' article titled '1811–12 New Madrid Earthquakes: An NEO Connection?' [33] discusses the possibility that the New Madrid Earthquake of 1811 may be related to the Great Comet of that year. Most of the article focuses on the possibility of finding an impact zone related to these events. The evidence seems inconclusive, but based on what I've read, this may be an incorrect approach to understanding the historical data, since comets do have the tenancy to explode in the atmosphere causing much havoc on the ground. In fact, there appears to be eyewitness testimony from people who lived through this event of a falling charcoal-like substance as well as visible undulations of the earth itself. An eyewitness report:

> ... the earth was observed to be as it were rolling in waves of a few feet in height, with visible depressions between. By and by these waves or swells were seen to burst, throwing up large volumes of water, sand and a species of charcoal...[34]

This evidence may imply an overhead explosion of sorts, which could have also set off some of the earthquakes in this region.

---

[33] http://www.datasync.com/~rsf1/1811.htm

[34] http://www.showme.net/~fkeller/quake/lib/eyewitness1.htm

Another point that should be made is *the reference to human conflict* during this time period. Obviously the War of 1812 between the American republic and Britain stands out as one example. Another would be Napoleon's march into Russia and his eventual defeat during an abnormally cold winter. Fritzius draws attention to the following statement by John Kezys ('Napoleon's Comets', *Orbit*, February 1996):

> As Napoleon marched into Russia with an army of seven hundred thousand strong, the Great Comet (of 1811) developed a tail one hundred million miles long. Following initial victories Napoleon overextended himself. After the invasion of Moscow he ran short of supplies and the winter proved unforgiving. Hundreds of thousands died while the comet performed frightening acrobatics by splitting in two.

This may go along with Victor Clube's analysis of the association between comets and turbulence within human civilization.

■ **1823 (November 10)** – Waseda, Japan: Meteorite strikes house.

■ **1825 (January 16)** – Oriang, India: Man reported killed, woman injured by meteorite fall.

■ **1827 (February 27)** – Mhow, India: Man struck on arm, tree broken by meteorite.

■ **1835 (November 14)** – Belley, France: Fireball sets fire to barn.

■ **1836 (December 11)** – Macaé, Brazil: Several homes damaged, several oxen killed by meteorite.

■ **1841** – Chiloe Archipel, Chile: Fire caused by meteorite fall.

■ **1845–6** – Ch'ang-shou, Szechwan, China: Stone meteorite damages more than 100 tombs.

■ **1847 (July 14)** – Braunau, Bohemia: A 37-lb iron smashes through roof of house.

■ **1850 (October 17)** – Szu-mao, China: Meteorite falls through roof of house.

■ **1858 (December 9)** – Ausson, France: Building hit by meteorite.

■ 1860 (May 1) – New Concord, Ohio, U.S.: Colt struck and killed by meteorite.

■ 1868 (August 8) – Pillistfer, Estonia: Building struck.

■ 1869 (January 1) – Hessle, Sweden: Man missed by few meters.

■ 1870 (January 23) – Nedagolla, India: Man stunned by meteorite. (Don't know if this means the man was 'amazed' or if he was hit and physically knocked senseless.)

■ 1871 (October 8) – Great Chicago Fire, U.S. (Another item that John Lewis has not entered into his calculations.)

■ 1872 – Banbury, U.K.: Fireball fells trees, wall.

■ 1874 (June 30) – Chin-kuei Shan, Ming-tung Li, China: Thunderstorm; huge stone fell, crushed cottage, killed child.

■ 1876 (February 16) – Judesegeri, India: Water tank struck by meteorite.

■ 1877 (January 3) – Warrenton, Missouri, U.S.: Man missed by few meters.

■ 1877 (January 21) – De Cewsville, Ontario, Canada: Man missed by few meters.

■ 1879 (January 14) – Newtown, Indiana, U.S.: Leonidas Grover reported killed in bed by meteorite (possible hoax in *Paducah Daily News*).

■ 1879 (January 31) – Dun-le-Poelier, France: Farmer reported killed by meteorite.

■ 1879 (November 12) – Huan-hsiang, China: Rain of stones; many houses damaged; sulfur smell.

■ 1881 (November 19) – Grosliebenthal, Russia: Man reported injured by meteorite.

■ 1887 (March 19) – Barque J.P.A., North Atlantic: Fireball *"fell into water very close alongside."*

■ 1893 (November 22) – Zabrodii, Russia: Building struck by meteorite.

■ **1897 (March 11)** – New Martinsville, West Virginia, U.S.: A man was reportedly struck, a horse killed, and walls pierced.

■ **1906 (November 4)** – Diep River, South Africa: Building struck.

■ **1907 (September 5)** – Hsin-p'ai Wei, Weng-Li: Stone fell; whole family crushed to death.

■ **1907 (December 7)** – Bellefontaine, Ohio, U.S.: Meteorite starts fire, destroys house.

■ **1908 (June 30)** – Tunguska valley, Siberia: Two reportedly killed, many injured by Tunguska blast.

■ **1909 (May 29)** – Shepard, Texas, U.S.: Meteor drops through house.

■ **1910 (April 27)** – Mexico: Giant meteor bursts, falls in mountains, starts forest fire.

■ **1911 (June 16)** – Kilbourn, Wisconsin, U.S.: Meteorite struck barn.

■ **1911 (June 28)** – Nakhla, Egypt: Dog struck and killed by meteorite.

■ **1912 (July 19)** – Holbrook, Arizona, U.S.: Building struck; 14,000 stones fell; man missed by a few meters.

■ **1914 (January 9)** – Western France: Meteor explosions break windows.

■ **1914 (November 22)** – Batavia, New York, U.S.: Meteorites damage farm.

■ **1916 (January 18)** – Baxter, Missouri, U.S.: Building struck.

■ **1917 (December 3)** – Strathmore, Scotland: Building struck.

■ **1918 (June 30)** – Richardton, North Dakota, U.S.: Building struck.

■ **1921 (July 15)** – Berkshire Hills, Massachusetts, U.S.: Meteor starts fire in Berkshires.

■ **1921 (December 21)** – Beirut, Syria: Building hit.

■ 1922 (February 2) – Baldwyn, Mississippi: Man missed by 3 meters.

■ 1922 (April 24) – Barnegat, New Jersey, U.S.: Rocked buildings, shattered windows, clouds of noxious gas, overhead explosion of comet fragment.

■ 1922 (May 30) – Nagai, Japan: Person missed by several meters.

■ 1924 (July 6) – Johnstown, Colorado, U.S.: Man missed being hit by one meter.

■ 1927 (April 28) – Aba, Japan: Girl struck and injured by 'dubious' (?) meteorite.

■ 1929 (December 8) – Zvezvan, Yugoslavia: Meteor hits bridal party, kills one.

■ 1930 (August 13) – Brazil: The 'Rio Curaca Event' (Brazlilian 'Tunguska Event'). Fire and 'depopulation'; an ear-piercing 'whistling' sound, which might be understood as being a manifestation of the electrophonic phenomena which have been discussed in *WGN* over the past few years; the sun appearing to be 'blood-red' before the explosion. The event occurred at about 8:00 local time, so that the bolide probably came from the sunward side of the earth. If the object were spawning dust and meteoroids – i.e. was cometary in nature – then, since low-inclination eccentric orbits produce radiants close to the sun, it might be that the solar coloration (which, in this explanation, would have been witnessed elsewhere) was due to such dust in the line of sight to the sun. In short, the earth was within the tail of the small comet. There was a fall of fine ash prior to the explosion, which covered the surrounding vegetation with a blanket of white.

■ 1931 (July 10) – Malinta, Ohio, U.S.: Blast, crater, smell of sulfur, windows broken in farmhouse; four telephone poles snapped, wires down; overhead cometary fragment explosion.

■ 1931 (September 8) – Hagerstown, Maryland, U.S.: Meteor crashes through roof.

■ 1932 (August 4) – Sao Christovao, Brazil: Fall destroys warehouse roof.

■ 1932 (August 10) – Archie, Missouri, U.S.: Homestead struck, person missed by less than one meter.

■ 1933 (**February 24**) – Stratford, Texas, U.S.: Bright fireball, 4-lb metallic mass falls, grass burned.

■ 1933 (**August 8**) – Sioux Co., Nebraska, U.S.: Man missed by a few meters.

■ 1934 (**February 16**) – Texas, U.S.: Pilot swerves to avoid crash with fireball.

■ 1934 (**February 18**) – Seville, Spain: House struck, burned.

■ 1934 (**September 28**) – California, U.S.: Pilot escapes fireball shower (one assumes this means he performed evasive maneuvers).

■ 1935 (**August 11**) – Briggsdale, Colorado, U.S.: Man narrowly missed by meteorite.

■ 1935 (**December 11**) – British Guyana: 21:00 local time; Lat. 2 degrees 10-min North; Long: 59 degrees 10-min West; close to Marudi Mountain. A report from Serge A. Korff of the Bartol Research Foundation, Franklin Institute (Delaware, U.S.) suggested that the region of devastation might be greater than that involved in the Tunguska Event itself. Eyewitness accounts were in accord with a large meteoroid/small asteroid entry, with a body passing overhead accompanied by a terrific roar (presumably electrophonic effects), later concussions, and the sky being lit up like daylight. A local aircraft operator, Art Williams, reported seeing an area of forest more than 20 miles (32 kilometers) in extent which had been destroyed, and he later stated that the shattered jungle was elongated rather than circular, as occurred at Tunguska and would be expected from the air blast caused by an object entering away from the vertical (the most likely entry angle for all cosmic projectiles is 45 degrees).

■ 1936 (**March 14**) – Red Bank, New Jersey, U.S.: Meteorite fell through shed roof.

■ 1936 (**April 2**) – Yurtuk, USSR: Building struck.

■ 1936 (**October 19**) – Newfoundland, Canada: Fisherman's boat set on fire by meteorite.

■ 1938 (**March 31**) – Kasamatsu, Japan: Meteorite pierces roof of ship.

■ 1938 (**June 16**) – Pantar, Phillipines: Several buildings struck.

■ **1938** (**June 24**) – Chicora, Pennsylvania, U.S.: Cow struck and injured.

■ **1938** (**September 29**) – Benld, Illinois, U.S.: Garage and car struck by 4-lb stone.

■ **1941** (**July 10**) – Black Moshannon Park, Pennsylvania, U.S.: Person missed by one meter.

■ **1942** (**April 6**) – Pollen, Norway: Person missed by one meter.

■ **1940s** – Qatar: A crater, believed to have been created by the impact of a falling meteor, found near Dukhan. Sheikh Salman bin Jabor al-Thani, head of the astronomical department at Qatar Scientific Club, said the Club believed that the meteor had hit Qatar in the 1940s. The Club started a search for evidence three years ago because of stories of a 'falling star' told by people of that era. They used Google Earth in the search, and succeeded in locating five craters, which were just visible on the surface.

■ **1946** (**May 16**) – Santa Ana, Nuevo Leon: Meteorite destroys many houses, injures 28 people.

■ **1946** (**November 30**) – Colford, Gloucestershire, UK: Telephones knocked out, boy knocked off bicycle.

■ **1947** (**February 12**) – Sikhote Alin, Vladivostok: An iron meteorite that broke up only about five miles above the earth rained iron. It produced over 100 craters with the largest being around 85-feet in diameter. The strewn field covered an area of about one mile by a half mile. There were no fires or similar destruction like that found at Tunguska; shredded trees and broken branches mostly. A total of 23 tons of meteorites were recovered, and it's been estimated that its total mass was around 70 tons when it broke up.

■ **1949** (**September 21**) – Beddgelert, Wales, U.K: Building struck.

■ **1949** (**November 20**) – Kochi, Japan: Hot meteoritic stone enters house through window.

■ **1950** (**May 23**) – Madhipura, India: Building struck.

■ **1950** (**September 20**) – Murray, Kentucky, U.S.: Several buildings struck.

■ 1950 (**December 10**) – St. Louis, Missouri, U.S.: Car struck.

■ 1953 (**March 3**) – Pecklesheim, FRG: Person missed by several meters.

■ 1954 (**January 7**) – Dieppe, France: Meteorite-building explosion, smashed windows.

■ 1954 (**November 28**) – Sylacauga, Alabama, U.S.: Mrs. Annie Hodges struck by 4-kg meteorite that crashed through roof, destroyed radio, and left a serious bruise on her hip. It is considered the only documented/verified case of a person being hit by a meteorite.[35]

■ 1955 (**January 17**) – Kirkland, Washington, U.S.: Two irons break through amateur astronomer's observatory dome; one sets a fire.

■ 1956 (**February 29**) – Centerville, S. Dakota, U.S.: Building hit.

■ 1959 (**October 13**) – Hamlet, Indiana, U.S.: Building hit.

■ 1961 (**February 23**) – Ras Tanura, Saudi Arabia: Loading dock struck.

■ 1961 (**September 6**) – Bells, Texas, U.S.: Meteorite strikes rook of house.

■ 1962 (**April 26**) – Kiel, FRG: Building hit.

■ 1963 – Massachusetts, U.S.: Meteorite fell.

■ 1965 (**December 24**) – Barwell, U.K: Two buildings struck and a car struck.

■ 1967 (**July 11**) – Denver, Colorado, U.S.: Building struck.

■ 1968 (**April 12**) – Schenectady, New York, U.S.: House hit.

■ 1969 (**April 25**) – Bovedy, N. Ireland: Building hit.

■ 1969 (**August 7**) – Andreevka, USSR: Building hit.

■ 1969 (**September 16**) – Suchy Dul, Czechoslovakia: Building hit.

---

[35] http://www.oberlin.edu/faculty/bsimonso/group9.htm

- 1969 (**September 28**) – Murchison, Australia: Building hit.

- 1971 (**April 8**) – Wethersfield, Connecticut, U.S.: House struck by meteorite.

- 1971 (**August 2**) – Havero, Finland: Building hit.

- 1972 (**August 10**) – Utah, U.S./Alberta, Canada: The Great Daylight 1972 Fireball (or US19720810). An earth-grazer meteoroid passed within 57 km of the surface of the Earth at 20:29 UTC on August 10, 1972, or 1.01 Earth radii from the centre of the Earth. It entered the Earth's atmosphere in daylight over Utah, United States (14:30 local time) and passed northwards leaving the atmosphere over Alberta, Canada. It was seen by many people, recorded on film and by space-borne sensors. Analysis of its appearance and trajectory showed it was a meteoroid about two to ten meters in diameter, in the Apollo asteroid class, in an orbit that would make a subsequent close approach to Earth in August 1997. In 1994 Zdenek Ceplecha re-analysed the data and suggested the passage would have reduced the meteoroid's mass to about a third or half of its original mass. The meteoroid's 100-second passage through the atmosphere reduced its velocity by about 800 meters per second, and the whole encounter significantly changed its orbital inclination from 15 degrees to 8 degrees.

- 1973 (**March 15**) – San Juan Capistrano, California, U.S.: Building hit.

- 1973 (**October 27**) – Canon City, Colorado, U.S.: Building hit.

- 1974 (**August 18**) – Naragh, Iran: Building hit.

- 1977 (**January 31**) – Louisville, Kentucky, U.S.: Three buildings and a car struck.

- 1979 (**June 7**) – Cilimus, Indonesia: Meteorite fell in garden.

- 1979 (**September 22**) – The Vela Incident (sometimes known as the South Atlantic Flash): The flash was detected on 22 September 1979, at 00:53 GMT, by a U.S. Vela satellite that was specifically developed to detect nuclear explosions. The satellite reported the characteristic double flash (a very fast and very bright flash, then a longer and less-bright one) of an atmospheric nuclear explosion of two to three kilotons, in the Indian Ocean between Bouvet Island and the Prince Edward Islands at 47° S 40° E. Hydrophones operated by the U.S. Navy detected a signal which was consistent with a small nuclear explosion on or slightly under the surface of the water near the Prince Edward Islands. The radio telescope at Arecibo, Puerto Rico, also detected an anomalous traveling ionospheric disturbance at the same time. *"There remains uncertainty about whether the South Atlantic Flash in September 1979 recorded by optical sensors on the U.S. Vela satellite was a nuclear detonation and, if so, to whom it belonged."*

■ **1981** (**June 13**) – Salem, Oregon, U.S.: Building hit.

■ **1982** (**November 8**) – Wethersfield, Connecticut, U.S.: Pierced roof of house.

■ **1984** (**June 15**) – Nantong, PRC: Man missed by seven meters.

■ **1984** (**June 30**) – Aomori, Japan: Building struck.

■ **1984** (**August 22**) –Tomiya, Japan: Two buildings hit.

■ **1984** (**September 30**) – Binnigup, Australia: Two sunbathers missed by 5 meters.

■ **1984** (**December 5**) – Cuneo, Italy: Strong explosion, building flash; windows broken; daytime fireball 'bright as Sun'.

■ **1984** (**December 10**) – Claxton, Georgia, U.S.: Mailbox destroyed by meteorite.

■ **1985** (**January 6**) – La Criolla, Argentina: Farmhouse roof pierced, door smashed; 9.5 kg stone misses woman by 2 meters.

■ **1985** (**June 26**) – Hartford, Connecticut, U.S.: A 1,500-pound slab of ice, six-feet long and eight-inches thick flattened a picket fence. The ground shook with the impact. A 13-year-old boy and his friend were standing ten feet away.

■ **1986** (**July 29**) – Kokubunji, Japan: Several buildings hit.

■ **1988** (**March 1**) – Trebbin, GDR: Greenhouse struck by meteorite.

■ **1988** (**May 18**) – Torino, Italy: Building struck.

■ **1989** (**June 12**) – Opotiki, New Zealand: Building hit.

■ 1989 (**August 15**) – Sixiangkou, PRC: Building hit.

■ 1990 (**April 7**) – Enschede, Netherlands: House hit by believed fragment of Midas.

■ 1990 (**July 2**) – Masvingo, Zimbabwe: Person missed by 5 meters.

■ 1991 – Tahara, Japan: Meteorite struck deck of car-transport ship; made crater.

■ 1991 (**August 31**) – Noblesville, Indiana: Meteorite fall missed two boys by 3.5 meters.

■ 1992 (**August 14**) – Mbale, Uganda: Forty-eight stones fall; roofs damaged, boy struck on head.

■ 1992 (**October 9**) – Peerskill, New York, U.S.: Car trunk, floor, pierced by meteorite.

■ 1994 (**January 18**) – Cando, Spain: An explosion that occurred in the village of Cando, Spain, in the morning of 18 January 1994. There were no casualties in this incident, which has been described as being like a small Tunguska Event. Witnesses claim to have seen a fireball in the sky lasting for almost one minute.

A possible explosion site was established when a local resident called the University of Santiago de Compostela to report an unknown gouge in a hillside close to the village. Up to 200 square meters of terrain was missing and trees were found displaced 100 meters down the hill.

■ 1994 (**July 16**) – Fragments of Comet Shoemaker-Levy begin impacting Jupiter.

■ 1994 (**October 20**) – Coleman, Michigan: Meteorite penetrated roof of house.

■ 1995 – Neagari, Japan: Meteorite penetrated car trunk.

■ 1996 (**November 26**) – Honduras: According to the Associated Press, *"A meteorite slammed into a sparsely populated area of Honduras last month, terrifying residents and leaving a 165-foot-wide crater, scientists confirmed Sunday. Near San Luis, in the western province of Santa Barbara."*

■ 1997 (April 11) – Chambrey, France: Meteorite penetrated roof of car; set fire.

■ 1998 (June 13) – Portales, New Mexico: Meteorite penetrated barn roof.

■ 1998 (July 12) – Kitchener, Ontario, Canada: Meteorite falls one meter from golfer.

■ 2000 (January) –Whitehorse, Yukon, Canada: A 150-tonne meteoroid lit the skies over Whitehorse, and exploded over a lake about 100 kilometres south of the city. The Tagish Lake meteor produced a treasure of information about a rare kind of meteorite.

■ 2000 (January) – Iberian Peninsula: Ice chunks weighing up to 6.6 pounds rained on Spain for ten days, causing extensive damage to cars and an industrial storage facility. At first scientists thought the phenomenon was unique to Spain. During the past three years, however, they've accumulated strong evidence that megacryometeors are falling all around the globe. More than fifty falls have been confirmed, and researchers believe that is a small fraction of the actual number, since others may hit unoccupied areas or melt before discovery. Most megacrymeteor falls occur in January, February and March. Megacryometeors show the telltale onionskin layering seen in hailstones. They also contain the dust particles and air pockets found in hail. But they are formed in cloudless skies, a characteristic that defies research on hail formation.

■ 2001 (July 25 to September 23) – Kerala, India: Red rain sporadically fell, staining clothes with an appearance similar to that of blood. Yellow, green and black rain was also reported. The rains were the result of the atmospheric disintegration of a comet, according to a study conducted at the School of Pure and Applied Physics of the MG University by Dr Godfrey Louis and his student Santosh Kumar. The red rain cells were devoid of DNA, which suggests their extra-terrestrial origin. The findings published in the international journal *Astrophysics and Space Science* state that the cometery fragment contained dense collection of red cells.

■ 2002 (June 6) – Asteroid/comet explosion over the Mediterranean: Estimated at five to ten meters in diameter, it released a burst of energy comparable to the nuclear bomb dropped on Hiroshima, Japan.

■ 2002 (September 24) – Near Bodaibo, Irkutsk, Siberia: Eyewitness accounts of the 1:50 a.m. event reported a large luminous object falling to earth near

Bodiabo in Siberia. Hunters in the region have also reported the existence of a crater surrounded by burnt forest, suggesting that an impact event had occurred. The event was detected by nearby geophones as a moderate earthquake. The event was also detected by a U.S. anti-missile defense military satellite. Some attempts were made to define the magnitude of the explosion. U.S. military analysts calculated it was between 0.2 and 0.5 kilotons, while Russian physicist Andrey Olkhovatov estimates it at 4-5 kilotons. Information about the event appeared in the mass media and among scientists after only a week. Another report says it occurred on the 25th of September at 10:00 p.m.

■ **2004 (June)** – Auckland, New Zealand: Meteor crashes through roof of home, damages sofa. The meteorite was a four billion-year-old 1.3 kg rock. *"There was this huge bang and a cloud of dust and debris went through the front room. I thought a car had hit the house."*
In the only account in New Zealand of a meteorite crashing into a house, the chunk of space rock punched a hole through the roof of the Archers' home, bounced off their couch, ricocheted off the ceiling and back on to the couch before ending up on the floor.

■ **2004 (September 3)** – Antarctica: A small asteroid exploded in the stratosphere above Antarctica, depositing sufficient micron-sized dust particles to cause *"local cooling, and much speculation as to the possible effects on the ozone layer."*

■ **2004 (December 26)** – Southeast Asia: An undersea earthquake occurred at 00:58:53 UTC with an epicentre off the west coast of Sumatra, Indonesia. It has been said that the earthquake was caused by subduction and triggered a series of devastating tsunamis; however, a comet/asteroid strike cannot be ruled out. Along the coasts of most landmasses bordering the Indian Ocean, more than 225,000 people in eleven countries were killed. Coastal communities were inundated with waves up to 30 meters (100 feet) high. It was one of the deadliest natural disasters in history. Indonesia, Sri Lanka, India and Thailand were hardest hit. See: 'Mega-tsunami hit southeast Asia 700 years ago'[36] and compare the events of 1430 with those above.

■ **2006 (February 1)** – Calgary, Alberta, Canada: On 1 February, twenty people reported seeing a fireball, an exceptionally bright meteor, streak across the sky just before 7 a.m., lasting for several seconds before breaking up into

---

[36] http://www.sott.net/articles/show/168328

fragments. It was estimated that remnants of the meteorite landed about 400 km south of Calgary somewhere in Montana, about two minutes after it appeared as a ball of fire.

■ **2006 (February 1)** – Bangladesh: The Dhaka *Daily Star* newspaper published the following report:

> A 'meteor' from outer space fell with a big bang on a field in the Singpara village of Sadar Upazila yesterday afternoon, creating panic and curiosity among people. No one was reported hurt. Superintendent of Police Khandker Golam Farooq rushed to the spot and asked his companions and villagers to dig the earth near the house of one Fazlur Rahman from where smoke was still emitting.
>
> To their amazement they found a lead-like black material three feet below the earth. Hot and weighing 2.5 kg, the triangular material looked like a mortar shell, witnesses said. The meteor was kept in custody of the Thakurgaon Police Station.

■ **2006 (February 17 and 20)** – Scotland: The U.K.'s *Daily Record* reported:

> The hunt is on for the crash sites of two meteors near Stirling Castle. Scientists have been spurred into action by reports of spectacular 'balls of fire' falling in the area. If discovered, they would be the first meteorites confirmed to have hit north of the border for almost 100 years. The incidents, reported by several witnesses, were on the evenings of Friday, February 17 and the following Monday, February 20. ...
>
> John Faithfull, curator of mineralogy and petrology at Glasgow University's Hunterian Museum, said yesterday: 'Although meteorite falls are rare everywhere, Scotland seems to have escaped remarkably lightly. There have only been four meteorites recovered from Scotland, compared with more than eighteen from England and Wales. Statistically, we are overdue another one.'

■ **2006 (April 12)** – Australia: A Perth astronomer reported that a spectacular light show in the sky was a meteor. Sightings were made as far south as Albany and inland through the Wheat Belt. It lit up the countryside for hundreds of kilometers around the southwest of Western Australia. Witnesses say the sky lit up about 9:00 p.m. AEDT, and the light was followed by a thundering sound that shook buildings.

■ **2006 (May 4)** – Texas, U.S.: Astronomers said a large meteor shower crossed straight over El Paso just before 9:45 p.m. on May 4. One meteor was so large that it cast an orange glow against the mountain.

The animals were going wild, the horses were bucking and dogs were barking and howling and then, all of a sudden right above my house, there was a big bright light and then just 'Bang!' And it lit up the five acres that are around us, and then I covered my eyes like this because it was bright and when it got past I saw there was a tail and it just went 'shhhh' toward the Hueco Mountains.

■ **2006 (June 2)** – Minnesota/Wisconsin/North Dakota, U.S and Canada: A fireball was spotted estimated to be some twenty miles above the Earth's surface. A sonic boom was heard in the Lake of the Woods area of Minnesota, so there may be some pieces of the meteor that survived the fall.

■ **2006 (June 19)** – Pennsylvania, U.S.: Residents of the Tuscarawas Valley who heard a deafening boom about 12:40 a.m. on Monday the 19th and stepped outside likely saw what one person described as *"a marvelous fireball with red streaks in the sky."* It probably was a meteor falling through the atmosphere. Numerous callers reported a large red fireball. Several said their homes shook. New Philadelphia police said they received reports from several callers who witnessed the fireball or heard the boom. One woman described it as *"a blue light that lit up the sky and went down."* Police in Dover said multiple callers reported they heard a loud bang and something rattled their windows. Air Traffic Command in Washington, D.C. confirmed that Cleveland's control center was checking into a meteor shower that occurred within its air space.

■ **2006 (July 10)** – South Africa: An ice ball that landed in Douglasdale, South Africa, might be one of the first 'megacryometeors' recorded in Africa. The ice ball, which landed on the pavement in suburban Douglasdal, was about the size of a microwave oven. The impact of the ice ball's fall created a small crater on the pavement, which was covered with pieces of broken ice. Despite sharing many chemical characteristics with hail, ice balls are formed under clear-sky conditions. Ice balls have been recorded since the 19th century. They have the potential to damage people, buildings and cars, but no injuries were reported as a result of this one.

■ **2006 (July 14)** – Norway: At 10:20 a.m. a bus driver from Ås, south of Oslo, was sitting in the outhouse at his holiday cabin near Rygge on the 14th of July when he heard an enormous blast. Right after that, some particles from a meteor that exploded over the Oslo area rained down just outside. He said he didn't think too much about the surprising blast at first, dismissing it as probably coming from an exercise at a nearby military air station at Rygge. But he said the blast and the rumbling it caused was terrible. He was just hooking the door when he heard a new noise, a whistling sort of sound, followed by a new bang on some aluminum plates lying near the outhouse. Sure enough, it was particles from a meteor that exploded somewhere over the Oslo Fjord area on Friday morning.

Astronomers confirm Martinsen's remarkable discovery of meteorite particles on his property. *"This is Norway's 14th meteorite, but we've never heard about a meteorite landing so close to a person before."* A family from Moss, south of Oslo, came home from their summer holidays to find a meteorite in their garden. It's another remnant of the meteor that exploded over the Oslo Fjord area on the 14th of July. Astronomers in Norway are calling the discovery of meteorites around southeast Norway "incredible," and urge local residents to keep looking for more. *"Two branches on our plum tree were broken. I lifted them up and there lay this stone."* It had made a hole measuring about seven centimeters in his lawn.

■ **2006 (September 12)** – New Zealand: A small piece of rock found in a paddock in New Zealand may be a piece of the meteorite that streaked across the sky there on Tuesday the 12th, panicking residents who flooded emergency hotlines. A farmer found a 10 by 5 centimeter piece of 'almost weightless' rock in his field today near the town of Dunsandel, south of Christchurch. It was sent to New Zealand's National Radiation Laboratory for analysis. The meteorite tore across the sky over the northern half of the South Island in the afternoon, leaving a bright, burning trail behind it, and causing a sonic boom that rattled houses and shook the ground. It then apparently erupted into a fireball, sending forth a thick puff of smoke. People were sent running from the homes and offices when they heard the boom, fearing buildings could collapse. The sonic boom was registered on earthquake-detecting equipment. The boom meant the meteorite was probably travelling 'very low'. It was probably about the size of a basketball as it shredded through the sky and became a "terminal fireball" at a speed of about 40,000 kph. *"If this had happened at night, it would have lit up the whole countryside."*

■ **2006 (October 10)** – Bonn, Germany: A fire that destroyed a cottage near Bonn and injured a 77-year-old man was probably caused by a meteor, and witnesses saw an arc of blazing light in the sky. Burkhard Rick, a spokesman for the police in Siegburg east of Bonn, said the fire gutted the cottage and badly burnt the man's hands and face in the incident on October 10.

■ **2006 (November 17)** – The Moon: NASA reports that meteoroids are smashing into the Moon a lot more often than anyone expected. That's the tentative conclusion of Bill Cooke, head of NASA's Meteoroid Environment Office, after his team observed two Leonids hitting the Moon on November 17, 2006.

"We've now seen 11 and possibly 12 lunar impacts since we started monitoring the Moon one year ago," says Cooke. *"That's about* **four times more hits than our computer models predicted.***"*

■ 2007 (January) – Tampa, Florida, U.S.: A 200-pound chunk of ice streaked through the clear Florida sky and landed in the back seat of a really nice red Ford Mustang. The car was totaled.

■ 2007 (January 4) – New Jersey, U.S.: Authorities were trying to identify a mysterious metallic object that crashed through the roof of a house in eastern New Jersey. Nobody was injured when the golf-ball sized object, weighing nearly as much as a can of soup, struck the home and embedded itself in a wall Tuesday night.

Approximately 20 to 50 rock-like objects fall every day over the entire planet, said Carlton Pryor, a professor of astronomy at Rutgers University. *"It's not all that uncommon to have rocks rain down from heaven,"* said Pryor, who had not seen the object that struck the Monmouth County home. *"These are usually rocky or a mixture of rock and metal."*

■ 2007 (January 10) – Russia: A meteorite fell in January in the Altai Territory in southern Siberia and searchers found an extraterrestrial substance which could be meteorite fragments. *"We have collected about 50 samples, and vitreous threads (traces of comet substance) were discovered in the first of them using a microscope."* Local motorists and residents witnessed the impact of a fiery ball, which eventually ended in a loud sound resembling an explosion.

■ 2007 (January 24) – Virginia, U.S.: Giles County residents were a little shaken after a tremor-like event, others say they heard a loud 'thunder-like' sound. Virginia Tech researchers say they received several calls about a meteor sighting the same time of the tremors. The bizarre incident took place around 8:00 p.m. Researchers say the seismic station in Giles County did get a very short but intense seismic signal.

■ 2007 (January 31) – Turkey: Police were inundated with calls from scores of people from Didim to Bodrum after they heard a big bang and a flash of light across the skies. The flashing green, yellow and red lights were from a meteorite which crashed through the earth's atmosphere and landed in Yesilkent. A startled man revealed that the rock had smashed a hole in the ground at the Green Park Complex, at Yesilkent, narrowly missing him by ten meters. Police reported that people from Bodrum, Milas and Didim had heard a bang and seen the flashing light across the skies at about 5:30 p.m.

■ **2007 (February 4)** – Midwestern U.S.: Scores of people all over the Midwest and Upper Midwestern United States reported seeing flames and fiery explosions in the sky Sunday night. From southeastern Wisconsin to as far as Des Moines, Iowa and St. Louis, people reported seeing balls of fire, possibly meteors, streaking across the sky on Sunday night. *"We had a pilot reporting seeing a meteor."* Reports came from residents in central Missouri, Illinois, Kansas, Wisconsin and Minnesota.

■ **2007 (February 15)** – Ohio, U.S.: Something happened at around 9 p.m. that a lot of people heard. But nobody seems to have any idea what it was. 'It' was a loud bang, something loud enough to be heard all over the county, and strong enough to make small objects move in houses. Rumors range from an earthquake to a meteor strike, a sonic boom or something ice-related.

At least one scientist believes the meteor could be the answer. There's no evidence to suggest an earthquake could have caused the bang, especially not over the range specified.

One man said he saw a meteor with a relatively long trail, with red, green and gold coloration. It was headed east to west and lasted about three seconds; after it faded, the sonic boom washed over him. *"I saw it first. It was the most eerie, cool, scary, wonderful thing. You just see this dragon tail going across the sky. All of a sudden, everything goes boom."*

■ **2007 (February 22)** – Rajasthan, India: Three people were killed and four injured in a mysterious blast in a village in India's northern Rajasthan state Thursday that villagers claim was caused by a meteorite, news reports said. Residents of Banchola village in the Bundi district, about 200 kilometers south of Rajasthan capital Jaipur, said the victims were sitting with some iron scrap in an open field when an 'object' fell from the sky and hit them.

■ **2007 (February 23)** – Panama: Panamanian geologists found a meteorite at Rio Hato, a coastal town west of the capital Panama City. The meteorite fell onto Rio Hato's beach. The landing was witnessed by a security guard, who described it as a ball of fire crashing down from the sky onto the sand. The 4.2 kg red object, measuring 20 centimeters in diameter, was to be X-rayed for more details. The meteorite shows burn marks on its exterior, and appears to be mainly carbon-based, in contrast to most meteorites, which mainly contain iron.

■ **2007 (March 15)** – Ontario, Canada: What Richard Yip-Chuck saw fall into a farmer's field Sunday evening looked like a long, white ball with orange sparks shooting off the back. The Holland Landing resident was driving along

Highway 7 with his wife and sons when they saw what looked like a fireball plummet to earth.

■ 2007 (**March 29**) – New Zealand: Flaming debris of a possible meteor almost hit a plane. The pilots of a Chilean passenger jet reported seeing flaming debris fall past their aircraft as it approached the airport at Auckland, New Zealand. The captain *"made visual contact with incandescent fragments several kilometers away."* The pilots reported the near-miss to air traffic controllers, reportedly saying the noise of the debris breaking the sound barrier could be heard above the roar of his aircraft's engines.

■ 2007 (**May 10**) – Spain: A fireball was spotted across central Spain. Scientists think some fragments may have fallen to earth in the Ciudad Real area. The fireball fell across the center of the country with sightings in Cuenca, Toledo, Ciudad Real and Valladolid. Scientists believe it was a meteorite and say it is quite a normal phenomenon, possibly a fragment from a comet which fell from earth orbit.

■ 2007 (**May 14**) – Vermont, U.S.: Recorded as a 2.1 temblor on the Richter scale, a quake hit Hubbardton, Vermont at 4:10 a.m. One resident, who said he was wide awake at 4 a.m., said he not only felt the earthquake, he saw what caused it. He said he saw something in the sky to the northeast of Lake Hortonia. He believes he saw a meteorite and that's what triggered the earthquake. *"It was like a streak of fire. I've heard meteorites hit before and that was what it sounded like. It was no earthquake, it was a meteor."*

■ 2007 (**May 26**) – Woburn, Massachusetts, U.S.: A meteorite reportedly punched a hole through a warehouse roof.

■ 2007 (**June 7**) – Norway: A large meteorite struck in northern Norway, landing with an impact an astronomer compared to the atomic bomb used at Hiroshima. The meteorite appeared as a ball of fire just after 2 a.m., visible across several hundred miles in the sunlit summer sky above the Arctic Circle. *"I saw a brilliant flash of light in the sky, and this became a light with a tail of smoke. I heard the bang seven minutes later. It sounded like when you set off a solid charge of dynamite a kilometer away."* The meteor struck a mountainside in Reisadalen. The country's leading astronomer said he expects the meteor to prove to be the largest to hit Norway in modern times, even bigger than the 198-pound Alta meteorite of 1904. *"If the meteorite was as large as it seems to have been, we can compare it to the Hiroshima bomb. Of course the meteorite is not radioactive, but in explosive force we may be able to compare it to the bomb."*

■ 2007 (June 10) – Sri Lanka: The strange objects that lit the night skies on June 10 have now been confirmed as meteors. *"This is the first time that meteors of such magnitude have fallen in Sri Lanka. The shockwaves and vibrations have been heard throughout the country, from Galle to Puttalam."* A Senior Consultant believes that two large meteoroids entered the atmosphere, the larger one splitting into two and the smaller one into about 25 fragments. The loud explosions were some of the particles exploding, probably about 50 to 100 kilometers above the ground. In Kovinna, Andiambalama, at 9.05 p.m. on the 10th, a woman had noticed something unusual in the western sky. A bright light, almost as large as the full moon, appeared to be moving towards her in a wide arc. Alarmed by thoughts of terrorist air attacks, she called out to her neighbour. Together they watched fearfully as the glowing object drew closer, landed on the roof and vanished completely. A few minutes later the air vibrated with a loud explosion. The next day they discovered that parts of the asbestos sheets on the roof were charred and cracked. A few pieces of rock and sand were scattered around the damaged area. Similar incidents were reported around the country that night. Several people in areas such as Puttalam, Maho and Bingiriya also noted the appearance of the bright light in the sky, as well as the loud explosion. In Kimbulapitiya a woman watched a flaming object land on a house and heard the booming sounds soon afterwards. In Campbell Place, Dehiwala, the roofs of two buildings were damaged, and a loud noise was heard.

■ 2007 (July 6) – Colombia, South America: An incoming object broke apart in the lower atmosphere with a trio of ferocious explosions that shattered windows and shook the ground violently. Moments later, stones rained from the sky and pelted homes in the poor barrios surrounding the city. Some smashed through the roofs of homes. Recovered objects were chondritic (rocky) meteorite.

■ 2007 (July 26) – Iowa, U.S.: A Dubuque woman said she was lucky to be alive after a 50 pound chunk of white ice crashed through the roof of her home, landing about 15 feet away from where she was standing. She said it sounded like a bomb exploded when the massive ball of ice hit her roof. Other large chunks of ice fell from the sky in this northeast Iowa city, tearing through nearby trees. Dubuque had clear skies at the time the ice fell.

■ 2007 (August 1) – India: Hotipur (Sangrur) village near Khanauri hit the headlines when a meteorite fell in the, leaving many villagers baffled. The police took possession of the 8-centimeter meteorite and handed it over to a three-member team of the Geological Survey of India. Curious villagers queued up in the fields to see the 'heavenly object', while the farmer, who was the only witness to the fall of the 'fireball', said, *"I got scared of the big fireball that was coming my way…. I ran for cover as I felt that it will fall on me."* [May be a hoax.]

■ **2007 (August 11)** – California, U.S.: Representatives with the Sonora Police Department and both the Tuolumne and Calaveras County Sheriff's Departments say they fielded numerous calls early in the morning in regards to a 'loud boom', and 'structures shaking'. There were several calls from residents who reported seeing 'a blue light', just before the 'loud boom'. The incident reportedly occurred at 12:09 a.m. The Police Department notes that it also received a call from a resident in Tuolumne, in which a female reported seeing what she thought was fireworks, and then something spiraling over her house. Early indication from the law enforcement agencies is that the loud boom was somehow the result of a meteor shower.

■ **2007 (September 15)** – Peruvian Highlands: A meteorite blew a 13-meter-wide hole in Carancas. A local official, Marco Limache, said that *"boiling water started coming out of the crater, and particles of rock and cinders were found nearby"*, as *"fetid, noxious"* gases spewed from the crater. After the impact, villagers who had approached the impact site grew sick from a then-unexplained illness, with a wide array of symptoms. Two days later, Peruvian scientists confirmed that there had indeed been a meteorite strike, quieting widespread speculation that it may have been a geophysical rather than a celestial event. At that point, no further information on the cause of the mystery illness was known. The ground water in the local area is known to contain arsenic compounds, and the illness is now believed to have been caused by arsenic poisoning incurred when residents of the area inhaled the vapor of the boiling arsenic-contaminated water.[37] The meteorite's impact sent debris flying up to 820 feet away, with some material landing on the roof of the nearest home 390 feet from the crater.

■ **2007 (October 3)** – Minnesota, U.S.: People across the Twin Cities reported seeing a 'metallic' object or 'flaming ball' falling from the sky. Broadcasters and emergency dispatchers got hundreds of calls from people who saw the object traveling from the northeast to the southwest. Residents of Lyon County in far southwestern Minnesota reported a loud boom that might have been connected with the sightings in the Twin Cities. A man who lives near the town of Amiret says it shook his house and sounded like a sonic boom from an F-14 breaking the sound barrier at close range. Coincidentally, at the same time, drivers

---

[37] http://en.wikipedia.org/wiki/Carancas_impact_event

in the Twin Cities metro were dodging debris in the middle of Interstate 94. Some drivers said the debris fell from the sky shortly after 2:00 p.m. Wednesday.

■ **2007 (October 30)** – Romania: A military airplane was hit by 4 'UFOs' while at an altitude of 6,300 meters, causing damage to the aircraft. While meteorites were excluded as a possible cause, the reasoning was simply that *"no astronomical data confirms entry of such bodies in the air that day"*, as if all such entries are recorded.[38]

■ **2008 (January 31)** – Didim, Turkey: Police were inundated with calls from scores of people from Didim to Bodrum after they heard a big bang and a flash of light across the skies. A startled Abdullah Arıtürk revealed that the rock had smashed a hole in the ground at the Green Park Complex, at Ye ilkent, narrowly missing him by ten meters.

■ **2008 (February 19)** – Northwest U.S.: An apparent meteor streaked through the sky over the Pacific Northwest, drawing reports of bright lights and sonic booms in parts of Washington, Oregon and Idaho. Although a witness reported seeing the object strike the Earth in a remote part of Adams County, in southeast Washington, it still has not been found. People in Washington, Oregon, Idaho, Montana and British Columbia reported seeing the bright fireball streaking across the sky about 5:30 a.m. At least one person said the object exploded on impact in eastern Washington and another report from southeastern Washington said someone felt tremors from the blast.

■ **2008 (March 5)** – Ontario, Canada: The Physics and Astronomy Department at the University of Western Ontario has a network of all-sky cameras in Southern Ontario that scan the sky monitoring for meteors. Associate Professor Peter Brown, who specializes in the study of meteors and meteorites, says that Wednesday evening (March 5) at 10:59 p.m. EST these cameras captured video of a large fireball. The department also received a number of calls and emails from people who actually saw the light.

■ **2008 (March 8)** – Turkey: A resident of Yaka said he heard a loud roaring noise at around 11:20 a.m. sounding as if *"a plane had crashed."* *"We were amazed to find such a small stone after that thunderous sound. It was black and about 40 centimeters in diameter, weighing three kilograms at most,"* another said, adding that the meteorite opened a small crater in the ground and created a cloud of dust.

---

[38] http://www.sott.net/articles/show/160342-Romanian-airplane-hit-by-an-UFO

■ **2008 (March 10)** – Sudbury, Canada: Great balls of fire were seen falling from the sky. While most sightings were reported around 1:30 p.m. near Sudbury, Hagar, Highway 69 North and North Bay, Wayne Lachance spotted something in the sky earlier in the morning. Lachance was driving home to Massey after a night shift at Vale Inco Ltd. when something caught his eye around 7:30 a.m. *"I thought it was a real bright star,"* he said. *"It was getting brighter and coming down with sparks."* Lachance arrived home and looked outside his bedroom window to see 'spirals of smoke' falling.

■ **2008 (March 13)** – The Moon: Meteorite videotaped hitting the Moon.

■ **2008 (April 6)** – Argentina: A space rock reportedly crashed somewhere in Entre Rios Province, some 260 miles northwest of Buenos Aires. Milton Blumhagen, a witness and astronomy buff said: *"For three or four seconds I saw an object in flames, changing color until it turned blue when it approached the ground."* A fire department source said the impact was felt for miles around. No damage was reported.

■ **2008 (April 15, 16, 18)** – Illinois, U.S.: Maybe there was a comet-fragment impact or two (or three) over a period of several nights; perhaps a couple of overhead explosions and then, later, a ground impact. That would explain the booms, earthquake and lights in the sky spread out over three days. Read the following stories and judge for yourself:

Damage Control: Mysterious booms, lights over Indiana were just F-16s [39]

A sonic boom and fireballs and flaming debris that Kokomo-area residents reported seeing in the sky Wednesday night prompted Howard County's police agencies to conduct a two-hour search for what many residents thought was a crashed aircraft. As it turned out, the fireballs were flares fired by F-16s that are part of the 122nd Fighter Wing, an Indiana Air National Guard unit based at Fort Wayne International Airport. [...] Staff Sgt. Jeff Lowry, with Indiana National Guard's headquarters in Indianapolis, said the jets taking part in the training are not supposed to exceed the speed of sound, which is about 760 mph, because supersonic speeds produce sonic booms. He said the 122nd's commander, Col. Jeff Soldner, will investigate why at least one jet reached supersonic speeds Wednesday night over Howard and Tipton counties, and also on Tuesday night over the Logansport area, shaking the ground below. [...] He said F-16 training often involves

---

[39] http://www.sott.net/articles/show/154005

the aircraft dropping flares from more than 10,000 feet above the ground, a technique that can allow the jets to evade heat-seeking missiles in combat. [...] Logansport Police Chief A.J. Rozzi said he heard a loud sonic boom on Tuesday night, and then heard the sound of a jet high overhead. He said residents also reported seeing fire streaks in the sky. He said it is common for the 122nd to conduct missions in the area and believes F-16 training almost certainly explains the sights and sounds. "They've been doing that training for quite a while. I don't know what maneuvers they're actually doing, but they do shoot out streaks of light," he said.

5.4 earthquake rocks Illinois; felt 350 miles away [40]

A 5.4 earthquake that appeared to rival the strongest recorded in the region rocked people awake up to 350 miles away early Friday, surprising residents unaccustomed to such a powerful Midwest temblor. The quake just before 4:37 a.m. was centered 6 miles from West Salem, Illinois, and 66 miles from Evansville, Indiana.

It was felt in such distant cities as Chicago, Cincinnati and Milwaukee, 350 miles north of the epicenter, but there were no early reports of injuries or significant damage. [....] "You could hear a roaring sound and the whole motel shook, waking up the guests," Vibha Ambelal, manager of the Super 8 Motel in Mount Carmel, Illinois, near the epicenter, said in a telephone interview.

4.5 Magnitude Earthquake Hits Illinois, Continuing Series [41]

A 4.5-magnitude tremor struck southern Illinois on Monday continuing the series of aftershocks initiated by the 5.2 earthquake which hit the region Friday morning, the U.S. Geological Survey (USGS) informed. This was the 18th earthquake in that series and its epicenter was approximately six miles below ground and about 37 miles (60 km) north-northwest of Evansville, Indiana, or about 131 miles (211 km) east of St. Louis, the USGS revealed. [...] The 18 aftershock earthquakes which followed Friday's tremor haven't measured more than 3.9 on the Richter scale, but the first one was the biggest to shake the region, called the Illinois basin-Ozark dome, in over 40 years.

■ **2008 (April 16)** – Argentina: The Entre Ríos Astronomy Society in Argentina announced that on Wednesday, April 16th, 2008, at approximately 19:30 hours, they observed a highly luminous object that had all the characteristics of a bolide. This object was sighted from Paraná, Oro Verde and San Benito. According to witnesses, the bolide was intensely bright, with colours fluctuating

---

between green, yellow and red. It followed *a roughly north-east trajectory towards the south-west*, with an angle of 75 degrees. *One observer has stated that the bolide exploded before disappearing.*

It is not possible to discount the idea that this meteor relates to a similar fall which occurred the previous week over central Entre Ríos province, and which was observed across a wide part of Argentina. *The AEA also received over the past few days many reports of sightings of very luminous objects in different parts from the country*, e.g., from Mar del Plata, Tucumán, Zárate, Concordia, Ituzaingó (Prov. de Corrientes), etc.

■ **2008 (April 17)** – Argentina: [This may be the same event as reported on April 16th above.] A fireball fell somewhere in or nearby Entre Rios, 260 miles northwest of Buenos Aires.[42]

Mariano Peter from the Entrerriana Astronomy Association said there were reports from four witnesses. One of them described *"a strong light that passed at a high speed through the sky and at a low altitude, going towards the south and then fell in the distance."* Another witness said, *"It was very bright and it changed color between green and red."*

The first fireball was reported in Entre Rios on April 6th, 2008 [see above]. A witness said: *"For three or four seconds I saw an object in flames, changing color until it turned blue when it approached the ground."* A fire department source said the impact was felt for miles around. The next day a fragment of the space rock was recovered.

Smoke chokes Argentina's capital [43]

Smoke blanketed the Argentine capital Friday as *brush fires apparently set deliberately* consumed thousands of acres in the provinces of Buenos Aires and Entre Ríos. The smoke, from about 300 fires, is blamed for at least two fatal traffic accidents this week that left eight people dead. Sections of major highways and the Buenos Aires port, among the busiest in the world, have been closed. Incoming flights to the city's domestic airport, Jorge Newbery Airpark, have been diverted. *The Argentine government has blamed farmers looking to clear their land for crops and grazing* for the fires, which are estimated to cover 173,000 acres (70,000 hectares). "This is the largest fire of this kind that we've ever seen," Argentine Interior Minister Florencio Randazzo said Thursday. Randazzo called the situation a "disaster." As of Friday morning, little progress had been made extinguishing the blazes.

---

[42] http://www.jornadaonline.com/LeerNoticia.asp?Tabla=Noticias&Seccion=Nacional&id=11866

[43] http://www.sott.net/articles/show/154066

■ 2008 (April 20) – Russia:

Another overhead explosion? Two killed, 300 left homeless in Russian Far East fires [44]

Two people have died and 325 people including 18 children have been left homeless by fires that ripped through the Amur Region in Russia's Far East, local emergency services said.

The fires began on Sunday evening and continued until Monday morning in seven districts of the region. *Locals had set light to dry grass to free land for farming and other purposes, and the flames were spread by high winds,* a police source told RIA Novosti.

A total of 104 houses have been destroyed. One of those who died in the fires was a disabled man who was unable to leave his home. A total of 50 rescuers have been involved in the firefighting operation. People injured in the fires will receive 20,000 rubles ($900) in compensation, local authorities said. Over 11,000 hectares have been destroyed in an estimated *59 forest fires* currently burning in Russia's Far East, the Natural Resources Ministry said.

Curious how this report is similar to what happened several days ago in Argentina. And again, farmers all decided to set fires on the same day, and burning grass and high winds are blamed for the vast damage and considerable destruction. We wonder what kind of excuse authorities will invent when this kind of event happens in a non-agricultural area.

■ 2008 (April 29) – Willow Creek, California: A magnitude 5.2 earthquake 11 miles ease-southeast of Willow Creek was preceded by a sonic boom. Brenda Simmons, of SkyCrest Lake resort in Burnt Ranch, described it as *"a very loud noise [that] lasted 10 seconds max."*

■ 2008 (May 23) – Hoshiarpur, India: At 8:45 p.m., petrol pump staff watched a fireball fly through the sky and land in a nearby field, setting the bushes ablaze, after which they rushed to extinguish the fires. A month earlier, a similar object had reportedly fallen in Sangrur district, prompting a Geological Survey of India team to visit and examine the site.

■ 2008 (June 8) – Winsford, U.K: Graham Brooks reported a small meteorite impacted his garden at approximately 10:30 p.m. After hearing a large bang, he and his partner found a golf-ball-sized hole in the ground and a pile of rocky debris that caused an estimated £1,000 worth of damage to their caravan.

---

[44] http://www.sott.net/articles/show/154226

■ **2008 (July 1)** – San Bernardino Mountains, California: Dozens of witnesses across Southern California saw a glowing yellowish-green object with streaks of debris *"moving very fast across the northern sky"* and falling near the mountains. Perhaps not coincidentally, the 527,000-acre wildfires of 2008 had been burning since May 22, with several starting up to a month later (June 20), all of which were attributed to lightning and dry thunderstorms. One such fire was in the San Bernardino Mountains, but it is difficult to determine if the fire had been burning before the fireball sighting or not.

■ **2008 (July 31)** – Yellowknife, Canada: An unexplained explosion in the Northwest Territories left several whales dead on the shore under a cloud of black smoke.

■ **2008 (August 12)** – Orlando, Florida: Between 10 and 10:30 p.m., witnesses heard a loud explosion described as a sonic boom. The sound was followed by a fast-moving fire that destroyed at least four town homes.

■ **2008 (September 9)** – Eastern U.S.: NASA's SENTINEL all-sky camera picked up 25 bright meteors in a shower lasting 4 hours. Also known as the delta-Aurigids, the September Perseids come from an unknown comet and have been caught bursting only four times in the last century: 1936, 1986, 1994 and now 2008.

■ **2008 (September 25)** – Caltech scientist Paul Bellan was quoted as saying that *"The incidence of noctilucent clouds seems to be increasing ..."* These luminous clouds are coated with a thin film of sodium and iron, which collect in the upper atmosphere after being blasted off incoming micrometeors, suggesting that the increase of noctilucent clouds may be tied with an increase in micrometeors.

■ **2008 (October 7)** – Sudan: NASA scientists at Ames Research Center accurately predicted that a car-sized asteroid would enter the atmosphere 12 hours before it did, the first prediction of its kind. As it entered the Earth's atmosphere, the heat created a spectacular fireball, which was photographed, releasing huge amounts of energy as it disintegrated and exploded in the atmosphere, dozens of kilometers above ground. The asteroid exploded with the energy of around one kiloton, equal to the power of a small nuclear bomb. Remnants of the asteroid were discovered on December 8. Later analysis discovered the presence of amino acids.

■ **2008 (October 8)** – Pennsylvania, U.S.: At approximately 5:30 a.m. several pieces of ice crashed through the roof of Mary Ann and Perry Foster in their

York Township home. One of the pieces had hit Mary Ann's head. Six pounds of ice had created a hole 12 inches in diameter through their roof and ceiling.

■ **2008 (October 12)** – India: Over 200 trees were mysteriously felled after a deafening roar was heard near the Sindh Forest Range in Baltal. Earlier that morning, 79 residential houses, 58 cowsheds, eight shops and a local Masjid were reduced to ashes in mysterious fire, which broke out in Chui Draman village of Marwah tehsil. The fire was unexplained.

■ **2008 (October 15)** – Ontario, Canada: At 5:28 a.m., all seven cameras of Western University's Southern Ontario Meteor Network recorded a bright, slow fireball. As the Science Daily report put it: *"For the second time this year, The University of Western Ontario Meteor Group has captured incredibly rare video footage of a meteor falling to Earth. The team of astronomers suspects the fireball dropped meteorites in a region north of Guelph, Ontario, Canada, that may total as much as a few hundred grams in mass."*

■ **2008 (October 28)** – Kansas, Nebraska, Colorado, and Maryland, U.S.: All four states had reports of a comet in the evening, at approximately 7:00 p.m. Witnesses in Kansas reported that the object sped across the sky and was followed by a loud explosion. The meteor broke the sound barrier, producing a sonic boom that shook homes in Osborne. Cloudbait observatory in Colorado videotaped the fireball, and received over 100 reports from various states.

■ **2008 (November 13)** – Ohio, U.S.: After reports of a bright light with a trail of smoke and a loud crashing sound, authorities spent four hours searching for what they initially thought to be a crashed plane. After finding no wreckage, they concluded it was probably a meteor.

■ **2008 (November 17)** – Colorado, U.S.: NASA astronomers reported a Leonid meteor outburst of as many as 100 meteors per hour. As NASA reported, *"Almost no one expected the old stream to produce a very strong shower, but it did."*

■ **2008 (November 20)** – Alberta and Saskatchewan, Canada: After an enormous fireball was seen and videotaped streaking across the sky from Edmonton, university student Ellen Milley found the fist-sized fragments of the 10-tonne meteorite in a small pond 40 km from Lloydminster, Saskatchewan.

■ **2008 (December 6)** – Colorado: A superbolide 100 times brighter than the full moon was seen and videotaped at 1:06 a.m. Astronomer Chris Peterson, who photographed the event, said that *"In seven years of operation, this is the* *brightest fireball I've ever recorded.*" Radio astronomer Thomas Ashcraft also recorded the radio echoes produced by the superbolide's ion trail. University of Calgary researchers has since recovered over 100 meteorite fragments from the site, and speculates that thousands more exist, perhaps setting a new Canadian record.

■ **2008 (Summary)** – Worldwide: In addition to the above reports, the SOTT.net archives have reports of *67 fireballs* that were observed all over the world, including several that accompanied sonic booms; as well as *25 unexplained explosions or sonic booms*, one of which (Alabama, Aug. 18) accompanied a 2.6 magnitude earthquake.

■ **2009 (January 17)** – Sweden and Denmark: A spectacularly bright fireball lit up southern Sweden and eastern Denmark at 8:15 p.m. Over 400 people registered their reports with the official Danish meteor website. The event baffled Henning Haack, curator of the Geological Museum's meteorite collection, who told reporters: *"What was most unusual was the boom, together with the fact that it was so powerful. I've never personally experienced something like that in Denmark in the 10 years that I have been working with meteors."*

■ **2009 (January 31)** – India: A meteorite landed in open ground at 11:25 p.m. in the Akhnoor area in Jammu.

■ **2009 (February 24)** – Dallas, Texas: A six-pound chunk of charred metal blasted through the roof of a house and onto the ground floor of the kitchen.

■ **2009 (March 1–2)** – Savo, Finland: A fist-sized meteorite was photographed and probably landed along the border between Kangasniemi and Hankasalmi.

■ **2009 (March 29)** – East Coast, U.S.: Thousands of people from Maryland to Hampton Roads saw brilliant, streaking lights in the sky at 9:44 p.m. One or two minutes later, the display was followed by a loud sonic boom, which shook the ground and houses. The fireball(s), which were described as white, green and orange in color, also produced flashes of light and electrical problems. Joe Butler of Suffolk saw where it landed, telling reporters: *"The sky was light all of a sudden, like it was daytime. There it was, coming right at my car. It was so fast that I didn't even have time to think that I might have been in danger. It shot right over my car, it went down in the water right between the two bridges."*

■ **2009 (May 28)** – India: A 1-kilogram stony meteorite landed in the hamlet of Karimatti in the Hamirpur district at noon.

■ 2009 (June 1) – Mid-Atlantic Ocean: Air France Flight 447 en route from Rio de Janeiro to Paris broke up over the Mid-Atlantic, not having sent a mayday signal. The media provided conflicting reports of the cause (first focusing on weather then pilot tubes and instrumentation problems) and whether or not the flight was destroyed in flight. One report said that 30 recovered bodies were fully clothed and 'relatively well preserved', while earlier reports had said they had multiple fractures, and some were found with little to no clothing, suggesting a mid-flight breakup.

Hinting at a possible cause, a Spanish pilot reported seeing a flash of white light descend in the direction of the plane, which he was too far away to directly observe. An editorial for *Discover* asked the question: *"Did a meteor bring down Air France 447?"* If so, the shockwave and EMP produced by the explosion would account for the total failure of the plane's instrumentation, as well as the damage and mid-flight destruction.

■ 2009 (June 6) – Dhenkanal, India: Hemant Mohapatra was awakened from his sleep at around midnight to several 'stone pieces' blasting through his roof. One of the larger stones had hit Mr. Mohapatra and caused visible injuries to his body.

■ 2009 (June 11) – Germany: A pebble-sized meteorite struck 14-year-old Gerritt Blank on his way to school, leaving a 10-cm burn mark on the back of his left hand. He said, *"At first I just saw a large ball of light, and then I suddenly felt a pain in my hand. Then a split second after that there was an enormous bang like a crash of thunder. The noise that came after the flash of light was so loud that my ears were ringing for hours afterwards. When it hit me it knocked me flying and then was still going fast enough to bury itself into the road."* News reports claimed the meteorite had been traveling 30,000 mph.

■ 2009 (June 20) – Queensland, Australia: A fireball was observed to land on a mountain hear the town of Gin-Gin, setting nearby trees on fire. The fire, which was on the property of Mrs. Hazle Marland, was still burning 3 days later.

■ 2009 (July 20) – Jupiter: Fifteen years after Shoemaker-Levy 9, an object left an earth-sized impact on the surface of the planet. The discoverer of the impact, Anthony Wesley, was quoted as saying, *"If anything like that had hit the Earth it would have been curtains for us…"* Scientists at Sandia estimated the explosive yield to have been between three and five megatons. Astronomers believe the object was a few hundred meters in diameter. (Two lesser impacts, which did not produce visible debris, later occurred on June 3, 2010 and August 20, 2011.)

■ **2009 (September 3)** – Ireland: A huge 'sky explosion' occurred at approximately 9:00 p.m., prompting reports from west Cork, Kerry, Cavan and as far north as Donegal. An Irish astronomy group investigating the blast was quoted as saying: *"In the past two decades there have been two major explosions in the skies over Ireland."* One had been satellite wreckage, the other a meteorite which was later recovered.

■ **2009 (September 25)** – Grimsby, Ontario: Shortly after 9:00 p.m. an unusually bright fireball was observed and tracked by the University of Western Ontario's physics and astronomy department 100 kilometres above Guelph as the fireball streaked southeastward at a speed of about 75,000 km/h. Astrophysicist Doug Welch said, *"it's very rare for them to be this bright."* Initially, scientists were fairly certain that at least a small portion of the beachball-sized object had fallen near Grimby. About three weeks later, they found it. The golfball-sized fragment had smashed through the windshield of a Grimsby family's sport utility vehicle.

■ **2009 (September 28)** – Argentina: During the afternoon, a large fireball was seen by residents in Mendoza, La Pampa, San Luis, and Cordoba. It disintegrated with a large explosion before hitting the earth in a large, uninhabited field.

■ **2009 (October 8)** – Indonesia: At around 11:00 a.m., an asteroid detonated 15–20 km in the atmosphere above South Sulawesi, Indonesia, releasing about as much energy as 50,000 tons of TNT, according to a NASA estimate, about three times more powerful than the atomic bomb that leveled Hiroshima, making it one of the largest asteroid explosions ever observed.

The amount of energy released suggests the object was about 10 metres across, a size thought to hit Earth about once per decade. The blast was heard 10,000 miles away, and the fact that it was not spotted before entering the earth's atmosphere prompted fears that earth's defense against cosmic threats is woefully inadequate. However, the explosion received little to no coverage in the Western media.

■ **2009 (October 13)** – Belgium, Germany, Netherlands: A spectacular fireball display was photographed by Robert Mikaelyan showing the breakup of a fireball that was observed by hundreds of people from Belgium, Germany and the Netherlands. The bright, white fireball, with orange trail, broke into three large fragments (and several smaller ones), following which there was a rumble that shook houses. As with the Grimsby fireball, scientists were quoted as saying that a fireball of this size and brightness *"is likely seen anywhere in the world only every 20 to 25 years."*

■ **2009 (October 29)** – Latvia: After reports of a meteorite landing hear the Estonian border, scientists found a 27-foot diameter and 9-foot deep crater with smoking debris in its center. The rock would have had to have been at least 3 feet in diameter to cause a crater of this size. Unlike the Indonesian event, this one did receive plenty of news coverage. However, despite some of the original scientists who were on the scene professing it was a genuine crater, the last media reports quoted experts labeling it a hoax because the crater was *"too tidy"* and had fresh grass in it (as if the grass couldn't have been placed into the crater *after the fact*). We suspect this event was a genuine meteor impact, but that damage control went into overdrive to cover it up.

■ **2009 (November 14)** – Colorado, U.S.: A basketball-sized chunk of ice crashed through the roof of a family's Colorado home after apparently falling from an airplane passing overhead. Danelle Hagan and her 9-year-old daughter were at home in Brush when they heard the kitchen ceiling come crashing down. They were not injured. Investigators suspected it to be ice built up on an airplane. However, University of Denver astronomy professor Robert Stencel believes it may have been a megachryo meteorite, probably produced high in the atmosphere (although he does not completely rule out an extraterrestrial source).

■ **2009 (November 17)** – Kansas, U.S.: Chandler Harp, 10, was playing in the backyard of his Liberal home when he heard what sounded like an explosion about 15 feet from where he was standing. He looked over to see a plume of dirt and debris shoot 5 feet high. At the bottom of a foot-deep hole, he found a 2-inch rock. Biophysicist Don Stimpson confirmed the validity of the meteorite.

■ **2009 (November 18)** – Idaho and Utah, U.S.: A massive fireball that *"turned night into day"* over 500,000 square miles of land was observed all over Idaho and Utah, even as far as Las Vegas. In some areas, the flash of light was so bright it caused light sensor street lamps to shut off. In addition to the dramatic visuals, a rumble and sonic boom were heard as it detonated 100 miles mid-air with an energy equivalent to 0.5–1 kilotons of TNT. In the hours after, it left an iridescent blue dust cloud in the sky. Scientists speculated the object was between the size of a toaster oven and a washing machine. It prompted at least one witness to fear a nuclear strike.

■ **2009 (December 15)** – Maryland, U.S.: Early in the morning, Ocean City cab driver Derrick Miller saw a glowing hot object fall from the sky and land on the beach 20 yards away from him. The 20-gram, 1.5-inch long rock was red hot and had sparks coming out of two holes and had made a whole in the sand 18 inches wide and 6 deep.

■ **2009 (Summary)** – Worldwide: In addition to the above reports, the SOTT.net archives have reports of *114 fireballs* that were observed all over the world, including at least three 'extremely rare' daytime sightings, seven that accompanied sonic booms (several of which shook houses, and one of which accompanied a 3.5 magnitude earthquake), four extremely bright 'night to day' flashes, and three multiple fireball/fragmentation sightings. *26 unexplained explosions or sonic booms* were heard, one of which (Louisiana, Jan. 8) accompanied a bright flash, and another (South Korea, mid-April) was accompanied by flames and smoke, suggesting a possible impact.

■ **2010 (January 18)** – Eastern U.S.: Over 100 reports of a *"huge ball of bright white light"* was made in the eastern states, including Pennsylvania, Delaware, Maryland and Virginia. The mango-sized meteorite left behind a snaking smoke trail before punching a hole through the ceiling of a dentist's office in the Williamsburg Square Family Practice Office in Lorton, Virginia, at 5:45 p.m., narrowly missing patients and staff. Dr. Marc Gullani said, *"Literally an explosion went off."*

■ **2010 (February 11)** – Mexico: A meteorite crashed into Central Mexico, between Puebla and Hidalgo, at around 6:30 p.m., leaving a 30-meter crater and causing damage to a local road and bridge. The impact was so massive it shattered windows several kilometers away and swayed buildings. Soon after these reports, however, articles were posted denying the existence of the crater (which was previously reported as being cordoned off by military units), and ascribing the fireball to Russian space junk. Russian authorities denied the connection. The satellite's last orbital element was after the Mexico impact.

■ **2010 (March 29)** – Colorado, U.S.: A ten-pound rock landed just ten feet from the rural home of Roger Hebbert, on his birthday. His wife had heard a loud swooshing sound and felt a gust of wind travel through the house. It had created a funnel in the ground where it landed, and was surrounded by smaller rocks and holes.

■ **2010 (April 10)** – Ontario, Canada: Jack Steenhof was driving to Kitchener for a meeting when something shattered his sunroof. *"I thought someone shot at me"*, he told reporters. He was traveling under 10 km/h at the time, and found the inch-long meteorite on his roof after hearing it clatter (presumably between the roof and broken glass).

■ **2010 (April 14)** – U.S.: A giant green fireball seen over Indiana, Wisconsin, Iowa and Illinois was also seen to explode, producing a loud sonic boom just after 10:00 p.m. Other witnesses reported secondary explosions. The blast started 9 fires over a 1-mile stretch in southern Wisconsin. Scientists later found a 0.3 lb fragment with a fusion crust.

■ 2010 (**April 29**) – East Jakarta, Indonesia: Six months after the blast in South Sulawesi, a meteorite was believed to be responsible for an explosion that damaged three homes in Duren Sawit. A deep crater was found in one of the houses, with melted items and a residual heat footprint.

■ 2010 (**July 19**) – Bosnia: Radivoke Lajic's house has been hit by meteorites six times since November 2007. He has since reinforced his roof with steel to prevent any further damage.

■ 2010 (**July 21**) – Sussex, U.K.: Jan Marszel narrowly missed being struck by a meteorite. The five-centimeter object landed 5 yards away from Marszel, who was struck by one of the bouncing fragments.

■ 2010 (**August 8**) – Ohio, U.S.: Pat Foraker of Quaker City claimed he was hit in the shoulder by a meteorite. Foraker heard a whistling noise, after which the object hit him, cutting his shoulder, and landed in a swimming pool. The rock was still hot when he recovered it from the pool.

■ 2010 (**October 4**) – Central Java, Indonesia: Six months after the blast in East Jakarta, and one year after the one in South Sulawesi, another home was allegedly struck by a meteorite in Karanganyar, damaging its roof, kitchen and dining room. An egg-sized rock was recovered at the site, which had melted plastic and heated metal tableware.

■ 2010 (**Summary**) – Worldwide: In addition to the above reports, the sott.net archives have reports of *132 fireballs* that were observed all over the world and reported for the most part in major media. These included at least: 8 sightings of multiple fireballs or dramatic mid-air fragmentation; 11 sightings that were seen by multiple witnesses over several states or territories; 3 daytime fireballs; 3 'night to day' fireballs; 4 that accompanied explosions, booms and/or shockwaves, 2 of which resulted in fires. One fireball seen in Alabama on March 19 was described by experts as *"unusually low flying"*. *18 unexplained explosions or sonic booms* were heard. Also, there were *8 stories of meteorites* that were close enough to witnesses to be either heard or seen, and then recovered. One nearly hit a man in Tennessee, landing ten feet from him; one was found in a farmer's field in New Deal; one was seen and recovered by a young child in Texas; and one caused damage to a roof in Pennsylvania.

A flaming object seen over Oswego, Illinois, U.S., in March 2011.

■ 2011 (**April 1**) – Arizona, U.S.: A Southwest Airlines flight from Phoenix to Sacramento had to make an emergency landing in Yuma, AZ, after a three-foot hole 'opened up' in the roof of the plane. While the cause was not known, passengers reported hearing an explosion. At the end of the month, the FBI investigated another 'hole', this time in a plane at the Charlotte/Douglas International Airport in NC. They suspected a bullet hole. Small, sub-surface cracks were then found in three more Southwest Airlines planes in what Southwest's executive vice-president called *"a new and unknown issue."*

■ 2011 (**May 3–5**) – Croatia: Dumitru Zvanca was planting potatoes when a poolball-sized meteorite thought to be from the tale of Halley's Comet landed just inches away from him. Zvanca told reporters: *"I heard a brief whoosh of air and then something hit the ground just to one side of me with an enormous thud. I didn't see a meteor, but I saw the small crater of earth it made and whatever had hit the ground had sunk into the earth."* The round, black ball had buried itself 50–70 cm into the ground.

■ 2011 (**May 4**) – Poland: A 1-kg meteorite crashed through the roof of a farmhouse in the village of Soltmany.

■ 2011 (**May 6**) – New Jersey, U.S.: At around 11:35 a.m., a suspected meteorite landed in the front lawn of a Basking Ridge resident, digging a trench and spewing dirt over a 100-foot area of the lawn, driveway, and street.

■ 2011 (**June 4**) – Long Stratton, U.K.: Removal man James Barber heard a large bang as he returned to his 18-ton Mercedes Benz Axor lorry. He found a cell phone-sized hole that had been punched through the thick metal corner joint. While he did not find the object, brown soil-like specks were on the jagged edges of the cut metal.

■ 2011 (**July 16**) – Kenya: Residents of villages near Thika heard a loud explosion and witnessed a fireball that rained down several meteorites. Over 30 pounds of meteorites were recovered after the blast, including one fragment that weighed 11 pounds. A 70-gram fragment had punched a hole through a corrugated metal roof in Muguga, and one allegedly landed only five feet from a girl working in a coffee field.

■ 2011 (**Summer**) – France: An egg-sized meteorite smashed through the roof of the Comette family home outside of Paris some time during the summer, while the family was on holiday. Mineral expert Alain Carion called the event *"extremely rare"*, saying that France had had only 50 or so meteorites in the past 400 years.

■ 2011 (**September 1**) – U.S.: A report commissioned by NASA says the quantity of hazardous 'space junk' orbiting the earth in the form of meteoroids has reached a 'tipping point'. The increase in material was suggested to be man-made, and NASA did not raise the possibility that this increase could be due to a greater amount of cometary debris in our immediate environment.

■ 2011 (**August 13**) – California, U.S.: Mike Gibson of Sacramento was awakened by a loud boom to find a large 4.5–6-foot 'impact zone' on his roof, causing thousands of dollars of damage. The recovered fragment was just larger than a quarter.

■ 2011 (**September 27**) – Argentina: Just before 2 a.m. in Monte Grande, witnesses saw a blue fireball falling from the sky, followed by a large explosion. The explosion killed one woman, 43-year-old Silvina Espinoza, and injured nine others, destroying two houses, one business and several cars. In the days following, the government pinned the explosion on a pizza oven's poorly connected gas canister, despite the fact that chief of firefighters Guillermo Pérez had ruled out a gas-related incident and stated that the causes remained unknown. Additionally, a young man who claimed to photograph the fireball was arrested for giving false testimony and forced to 'admit' to hoaxing the photograph.

■ 2011 (**Summary**) – Worldwide: In addition to the above reports, the SOTT.net archives have reports of *109 fireballs* that were observed all over the world, 11 of which were visible over multiple states or territories (two were visible from six states each on January 11 and February 14), 10 of which had accompanying explosions or sonic booms, and 4 of which started forest or grass fires (in Croatia, Peru, Texas and India). There were 3 daytime sightings, 4 clusters/breakups, and 2 smoke/debris spirals. The May 20, Georgia sighting was of a human-sized meteor, the brightest on record for NASA's fireball-observing network and the multi-state February 14 daytime sighting was 5 tons, producing an explosion of approximately 100 tons of TNT. *24 unexplained ex-*

plosions or sonic booms were heard, 5 with probable fireball or 'flash' sightings, and 2 with accompanying earthquakes (2.7 in Georgia on November 9 and 1.7 in Philadelphia on May 28). Also, there were *8 stories of meteorites* that were close enough to witnesses to be either heard or seen, one of which produced a 3-foot deep hole in an Alabama backyard and shot up flames six feet high.

■ **2012 (February 27)** – NE U.S.: A green fireball was seen from New England and Canada between 10:00 and 10:30 p.m. One witness, Aidan Ulery of Exeter, found where it had landed, making a dent in the middle of the road. He said, "*I could smell it.*" He recovered several fragments. (This report was probably the same one seen by multiple witnesses throughout the northeastern U.S. and Canada but reported as taking place at 10:12 p.m. on the 28th.)

■ **2012 (February and April)** – U.S.: In response to the many accounts of fireballs in February, NASA called the phenomenon the 'Fireballs of February', as if it were a regular phenomenon. According to NASA's Bill Cooke, these fireballs are distinguished by their 'appearance and trajectory', and "*There is no common source for these fireballs, which is puzzling.*" Meteor expert Peter Brown cited contradictory research from the '60s, '70s and '80s that either confirmed or denied such a peak of 'February Fireballs'. One article suggesting such a peak was 'a 1990 study by Ian Halliday', where he also found peaks in late summer and fall. However, as one article on the subject said: "*The results are controversial, however. Even Halliday recognized some big statistical uncertainties in his results.*" And yet two months later, another previously unheard of phenomenon was introduced by the media and NASA: 'April Fireballs'. Bill Cooke was quoted as saying, "*There are two peaks: one around February and the other at the end of March and early April.*" Even Holliday, whose paper was cited as confirmation for 'February Fireballs', didn't find a peak in April. Seeing as how these terms were unheard of before 2012, it looks like NASA is reinventing history to make the increasing number of fireballs seem normal.

Meteor seen over the UK early April 2012.

■ **2012 (Summary)** – Worldwide: In five months, 2012 saw *40 fireball reports, 11 unexplained sonic booms, and 8 recovered meteorites.* The biggest meteorite fragment was found in China in February, it weighed 12.5 kg. The fireball reports contained a remarkable number of multi-state sightings – 10, one below the totals for both 2010 and 2011, including one in Australia which was called 'unprecedented' by experts. Seven exploded mid-air and/or were accompanied by sonic booms, including one in India that produced a 2.1 magnitude earthquake on May 22, and one daytime sighting in Nevada and California that was estimated to be the size of a minivan before breaking up, both of which fireballs left several meteorite fragments. 2012 also saw 3 daytime events, matching the yearly totals for 2009, 2010 and 2011.

■ **Other Data (Summaries):** The Cloudbait observatory in Colorado recorded 453 meteors from February 2004 to May 2012 (2004: 10, 2005: 19, 2006: 10, 2007: 57, 2008: 91, 2009: 98, 2010: 86, 2011: 55, 2012 [partial]: 27).[45] The American Meteor Society has collected American fireball reports from 2005 to the present. To give an idea of just how many fireball sightings there are just in the U.S. that aren't necessarily reported in the media (i.e., those that make up the bulk of the SOTT archives), here are their yearly totals: 466 (2005), 515 (2006), 537 (2007), 726 (2008), 693 (2009), 948 (2010), **1629** (2011), 795 (to June 6, 2012; by this time in 2011, there were just 568 reports).[46] 2011 and 2012 show an unprecedented number of sightings.

The last few years also saw an alarming number of 'close call' flybys from asteroids and meteoroids passing earth from within the distance of the moon, and often spotted with only days' notice. These occurred on 2009/03/02, 2009/11/06, 2010/01/13, 2010/04/09, 2010/09/08, 2010/10/12, 2011/02/11, 2011/03/16, 2011/06/02, 2011/06/27 (the fourth closest meteoroid on record, passing just 12,300 km from earth), 2011/11/08 (the fourth closest asteroid, i.e., greater than 50 meters in diameter, passing 324,900 km away), 2012/01/27 (one day's notice), 2012/04/01, 2012/05/13.

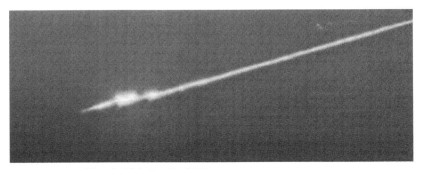

Meteor seen over Nevada, U.S., late April 2012.

---

[45] http://www.cloudbait.com/meteor/videos.php

[46] http://www.amsmeteors.org/fireballs/fireball-report/

PART III:

THE GOLDEN AGE, PSYCHOPATHY
AND THE SIXTH EXTINCTION

9. Creationism, Evolution and the Corruption of Science

In the chapters on the Hundred Years War, witches, and ancient spirituality, I discussed how the feminine energy of our society was debased step-by-step over millennia, as a consequence of planetary cataclysm beginning with the Great Deluge, also known as the Flood of Noah. But noticing what has happened is not the same as explaining why and how it could happen sociologically speaking. In those chapters, I sidestepped the idea of Atlantis to deal with other topics. One thing I want to do is address whether or not there was an advanced civilization known as Atlantis and if there was, what sociological processes might have been involved in its rise and subsequent fall, how mankind lost this knowledge, and whether these elements are related by a common factor that is present in our own civilization leading us in the same direction.

One of the first problems faced when dealing with this topic is the issue of the falsification of history and the corruption of scientific knowledge. I have already mentioned that the historical record of the past that we consider 'truth' nowadays was constructed at the end of a long period of planetary devastation (the Dark Ages followed by the Black Death) and this was done to restore order and control. The powers that be of the time, which is the same at about any time – religious and elite-class controllers – set about to deliberately establish a view of the past that placed themselves firmly in the seats of power, legitimating that authority in the eyes of the masses; masses that are true believers in what the elite tell them are much easier to manage. We can observe the same process taking place in our own day, one example being Christianity transformed from a religion of love, caring and peace to one of war and retribution. Actually, it just cycles around, because it was utilized similarly in 11th and 12th centuries which followed the Dark Ages, now strongly suspected to have been a time of cometary disaster in Europe. These cycles even prompt us to consider the possibility that there is a cause-effect relationship between periods of oppression of the masses by the elite and cosmic intervention that levels the playing field, so to say.

I've spent the past eight years or so buried in texts relating to astronomy, geology, paleontology, archaeology, sociology, anthropology, linguistics, genetics studies and more. The full truth about the corruption of science that I discovered there will be dealt with in another book. I've also spent years reading a mountain of revisionist history by authors such as Michael Cremo, Graham Hancock, Robert Bauval, Knight & Lomas, Andrew Collins, and many others. (This short list is not a recommendation and the absence of any other name is not a dismissal of their work – just the names that come immediately to mind to describe the genre. I do particularly like Andrew Collins and Knight & Lomas.) I would like to note that the word 'revisionism' somehow carries a negative connotation, but the fact is that constant revision of history as new data comes to light is a normal part of the scholarly process.

What I found – or realized – was that it would certainly be a very good idea if the revisionists were well trained in the scientific method, because they most certainly could make their cases tighter without so many careless mistakes or baseless assumptions. But, on the other hand, as for the work of the 'credentialed' scientists, I found it to not be credible at all. Despite the superior methodology, financial backing, institutional and peer support, much of it descends into nonsense when one realizes what is being omitted in order to support the a priori premises on which are built such elaborate, circular cognitive edifices.

So, let's get started. There are some indispensable concepts that I need to get out of the way before we get to the fun part: speculation on the available data.

Alexander Zinoviev, formerly of Moscow State University, was a logician and sociologist, winner of the 1982 Alexis Tocqueville prize for sociology, and member of many eminent academic societies and organizations, i.e., pretty mainstream, but, as we shall see, he obviously wasn't happy with the way science was being practiced. He wrote the following in the Foreword to mathematician Anatoly Fomenko's seminal work, *History: Fiction or Science?*:

> … what A.T. Fomenko and his colleagues had learnt over the course of their research was the fact that the entire history of humanity up until the XVII century is a *forgery of global proportions* … a falsification as *deliberate* as it is *universal*. …
> The first global falsification of history as discovered and brilliantly related by Fomenko was based on an erroneous temporal and spatial coordinate system of chronological events … [emphasis in original] (Fomenko 2006, xv–xvi)

It is unfortunate that neither Zinoviev nor Fomenko have factored in historical or social discontinuities due to cataclysmic disruption, which can make it so easy for ruthless individuals to seize power and re-create the past according to the needs of the present.

As I have already mentioned, the falsification of history that Fomenko has clearly identified occurred at the end of 300 years of disaster piled on disaster beginning with the decimation of the population of Europe by the Black Death. This certainly fits Zinoviev's model that, at a certain point, the old, distorted representation of history no longer serves its purpose and a new legitimation of authority is required to restore peace and keep the masses under control. What Fomenko notes is that the temporal and spatial parameters of much later times were imposed on the stories of the more distant past, but that does not entirely invalidate those stories; it just means that things were re-established and what was familiar to the 'trained specialists' was utilized to give the new view of the past a more realistic 'feel'.

In the following remarks, Zinoviev is talking about the ongoing falsification of history that is active today, here and now – and notice that he designates the theory of evolution as the breakpoint of this process, thereafter to be utilized as the standard-model framework:

> My sociological research of the great evolutionary breakpoint demonstrated that a new, blatant, global and premeditated falsification was already in full swing. Prior to becoming familiar with the writings of Fomenko, I had already known that the falsification of the past was a rather common phenomenon inherent in human existence. However, I was neither aware of the scale of this fraud as described by Fomenko and his fellow scholars, nor of its social type. My assumption had been that the blatant falsification of history on a planetary scale that I had discovered was the first one ... Let us call it the second falsification of the same variety. It differs from the first [discussed by Fomenko] in terms of pertaining to a different epoch. Its main subject is modern history ... One has to differentiate between the two kinds of falsification, the first one being the involuntary routine falsification of minor details that results from the mechanisms of gnosis ... or the entropy inherent in the framework of humanity's historical memory. The second is the extraordinary, premeditated and complex falsification that has distinct social causes. [emphasis in original] (Fomenko 2006, xv)

Like Zinoviev, I had also become aware of the falsification of history and I also thought it was just an accumulation of innocent errors.

> Even if we are to suppose that all those who partake in the creation of historical records see veracity as their mission, the result of their collective efforts is often the rendition of their own subjective views on history as opposed to what happened in reality. As centuries pass by, the stream of disinformation is fed by various sources and tributaries, which, in their multitude, produce the effect of impartial falsification of historical events. This stream also feeds on murky rivulets of countless liars and swindlers.

The false model of history serves its function for a certain while. However, humanity eventually enters a period when this distorted representation loses efficacy and stops serving its ends. This is where people are supposed to start searching for explanations and set out on their quest for a "truth". (Fomenko 2006, xvi)

In short, Zinoviev is talking about exactly what has happened in our own recent history and before we continue with his conclusions about it, let us take a quick look at something I wrote in my book, *Secret History*, about COIN-TELPRO:

Richard Dolan's *UFOs and the National Security State* is the first comprehensive study of the past 50 years of the U.S. government's response to the intrusion of UFO phenomena in America. The compiled evidence – which includes government documents – suggests that a group of specialists working in the shadows, set up and executed the most massive cover-up in the history of government; and that the Human Potential Movement and the subsequent New Age movements, were key elements of this cover-up. In other words, they not only have used the 'colorful community' of alternative ideas as an unwitting tool of disinformation, it is highly probable that most of it was literally created by them as COINTEL-PRO. According to analysts, cointelpro was the FBI's secret program to undermine the popular upsurge, which swept the country during the 1960s. Though the name stands for 'Counterintelligence Program', the targets were not enemy spies. The FBI set out to eliminate 'radical' political opposition inside the U.S. What a lot of people do not realize is that this was a high-level psychological operation specifically set up to vector 'ideological' trends – beliefs ...

What a lot of people don't keep in mind is the fact that cointelpro also concentrated on creating bogus organizations. These bogus groups could serve many functions which might include attacking and/or disrupting bona fide groups, or even just simply creating a diversion with clever propaganda in order to attract members away so as to involve them with time-wasting activity designed to prevent them from doing anything useful. COINTELPRO was also famous for instigation of hostile actions through third parties ...

Now, let us take a few logical steps. The UFO problem emerged into the national consciousness in 1947, or thereabouts. Not long afterward, a lot of people began asking a lot of questions. The government wasn't answering, and so the people began to band together to find out the answers for themselves. They started forming groups. And this is where things get just a bit curious. The thing that was most threatened by the UFO/alien issue seems to have been the standard monotheistic religions. Religion seems to be a necessary component of political control. Social control – that is the mainstay of religion – was most definitely under threat. In fact, what seems to be true is that it is not even clear that religions – as we know them – would have survived a full disclosure. So the logical conclusion is that part of the main reason for the cover-up was to 'protect the religious status quo'.

As things stood at the time, protecting the religious status quo – mainly the social controls that stem from religion – was iffy at best. After a century of scholarly investigation into many religious texts, and the raising of many questions about the 'old time religion', there were a lot of people in society who were most definitely turning away from religious dogma. It's fairly simple to take the next logical step and see that a combining of the questions of those who were disenchanted with religion, with the questions of those who wanted to know just what the heck was going on in terms of possible 'extraterrestrials', was seen as a dangerous and explosive mixture. Something had to be done.

The activities of COINTELPRO in attempting to neutralize political opposition have been pretty well exposed. But we are now considering the fact that, in addition to political activists, it seems that COINTELPRO has particularly targeted groups that are seeking the truth about the interactions between the U.S. government and ultraterrestrials, or so-called 'aliens'. That a long-time cover-up of these matters has been in effect is certainly evident to any careful researcher.

The COINTELPRO files show the U.S. government targeted a very broad range of religious, labor and community groups opposed to any of its agendas, and it is only logical to assume that the same type of operation would be created to cover up the 'alien agenda'. Such a theoretical COINTELPRO operation also goes far in explaining why, when the sincere researcher of UFO phenomena enters this field, he or she discovers only lies, lies, and more lies; confusion and disinformation. That is most definitely the signature of COINTELPRO.

What all of this seems to suggest is that the powers that be (ptb) have developed COINTELPRO to an all-new level of social shaping, cultural brainwashing, and the main targets of this activity would include virtually anyone who is seeking the truth about the shifting realities of our world. The cases of COINTELPRO activities against political groups must be no more than the tip of the iceberg, given that the great bulk of COINTELPRO-type operations remain secret until long after their damage has been done. By all indications, domestic covert operations have become a permanent feature of U.S. politics and social programming, and it is hardly likely, considering the evidence, that the New Age and Human Potential fields are exempt.

The implications of this are truly alarming. Those who manage to get close to the truth of these matters, despite the many obstacles in their path, face national covert campaigns to discredit and disrupt their research and reputations. Clearly, COINTELPRO and similar operations under other names also work to distort academic and popular perceptions of the problems facing our world. They have done enormous damage to the search for the Truth.

Now, let's get back to what Zinoviev wrote, with the above model in mind for comparison:

> ... One becomes aware of the necessity to update our view of the past in accordance with whatever the present stipulates. This awareness is the kind of craving that can only be satisfied by a "bona fide rectification" of history, which has to occur as a grandiose paradigm shift – moreover, it has to be a *large-scale organized operation*; one that shall result in an epochal falsification of the *entire history of humankind*. The issue at hand is by no means the falsification of individual observations of historical events, but rather the revision of the entirety of historical records describing the events which cannot be observed as a principle since they belong to the past. ... Trained specialists are a sine qua non for this – people whose activity shall have to be organized in such a manner that their collective output will result in the creation of a coordinated historical Gestalt. What they really have to do is create exactly the kind of past that is needed for the present, making use of whatever available material presents itself.
>
> ... The more recent and ongoing ... global falsification of history is based on a system of erroneous pseudoscientific sociological concepts based upon ideology and aided greatly by the modern information manipulation technology. ... Fomenko's works describe the technology of building a false model of human history which uses the art of manipulating the temporal and spatial coordinates of events. Many thousands of specialists in false historical models are already working on this second falsification – their forte is the ability to misrepresent historical events while giving correct temporal and spatial coordinates and representing individual facts veraciously and in full detail. The actual falsification is achieved via the selection of facts, their combination and interpretation, as well as the context of ideological conceptions, propagandist texts that they are immersed into, etc. [emphasis in original] (Fomenko 2006, xvi–xvii)

That is a grim indictment on science and the academic community, but is it true? I've spent eight years immersed in this material, pulling on threads, following references back to their point of origin, reading archaeological site reports, entire books that analyze the techniques of chipping stone and what that might reveal about the stone chippers, and more. At the end of it, I can only say that Zinoviev is correct. The question arises immediately: what kind of people would do this sort of thing? What kind of people would collude to mislead humanity so egregiously?

Nachman ben-Yehuda is another scientist who has taken note of this phenomenon and undertook to document it in detail in his exposé of the archaeology of Masada: *Sacrificing Truth: Archaeology and the Myth of Masada*. It's like a contagion, a spreading disease across all levels of society, and it seems that it *originates in pseudo-science that has taken over the field of science,* insinuating a completely different content of meaning under what were formerly clear scientific terms and conditions. This has infected all fields of research exactly as Zinoviev has said.

Let me make it clear that I do think the scientific method of cognition is *one* of the best tools we have for understanding our reality, but I have to say that the way it is wielded in mainstream science reveals that it has become more or less doublespeak: a term that originally meant one thing but whose meaning has gradually been corrupted. When you hear it bandied about nowadays, you can be pretty certain that its user is not being the least bit scientific in the original sense of the word but is rather a true believer in the religion of science.

Now, let's come back to the fact that Zinoviev mentions specifically the 'evolutionary breakpoint' – that is, the moment in our history when the theory of evolution entered the stage of history and became the ideological underpinning for the development of pseudo-science. (Here, the book to read is William R. Fix's *The Bone Peddlers: Selling Evolution,* on which the following discussion is based.)

On the *CBS Evening News* of 6 January 1981, Dan Rather recounted how, during the presidential campaign the previous autumn, Ronald Reagan had told an audience in Texas that he had *"a great many questions about evolutionary theory"* and further asserted that the theory *"is not believed in the scientific community to be as infallible as it once was believed."* Obviously, Reagan was courting the Bible-Belt voters. But, since Reagan had presumed to speak for the scientific community, it was not a surprise when said 'community' wanted to respond and this was the item that Dan Rather was reporting.

The spokesmen for the American Association for the Advancement of Science (AAAS) characterized Reagan's statement as *"tremendously unfortunate."* One scientist stated categorically that the *"100 million fossils identified and dated in the world's museums constitute 100 million facts that prove evolution beyond any doubt whatever."* Science historian William Fix examines the fossil evidence with some care and this turns out not to be true at all. In fact, there is so much evidence of falsification, fraud, and irrational belief among paleoanthropologists that one wonders why some of these 'eminent scientists' haven't been run out of town on a rail, tarred and feathered, or even – in some cases – arrested and tried for fraud.

What was extraordinary about the AAAS statement was not just that they characterized the statement of the President-elect of the United States as *"tremendously unfortunate"* but that they made the ridiculous claim that *"100 million fossils constitute 100 million facts that prove evolution",* a claim that Fix demolishes completely.

In March of the same year, 1981, a Judge Irving Perluss heard the case of one *Kelly Segraves vs the State of California*. Segraves, director of the Creation-Science Research Center in San Diego, sued claiming violation of the religious freedom rights of his children in California schools. The claim was that since California school board policy does not allow the 'divine origin theory' to be taught in biology, this was a violation of principle. The judge ruled that the rights of Segraves' children had not been violated but – and this is a big one – the judge ordered statewide distribution of an earlier school board statement cautioning that textbooks should avoid dogmatism on the question of man's origin. In short, textbooks could not claim evolution as an established fact, but must make it clear that it is only a theory.

This trial made big news (which is interesting in itself) and *Time* magazine headlined it: 'Darwin Goes On Trial Again' (16 March 1981). The *Time* article noted that, over the previous several years, apparently fundamentalism was on the rise and pressure from such groups had already forced changes in biology textbooks. In one such text, the discussion of the origins of life had been cut from 2000 words in the 1974 edition to 332 words in the 1977 edition. This sort of thing is more and more common in our current day. What is important here is that we see in these events the lines being drawn at the same point in time that several other interesting things were happening. The timing is so exact that I was baffled when I noticed it. It relates to something I wrote about in my book, *Secret History*:

> What strikes me as an essential turning point in this COINTELPRO operation was the beginning of the 'exposé' of two particular items that hold sway in certain 'conspiracy' circles to this very day: alien abduction and Satanic Ritual Abuse.
>
> The Gray-alien scenario was 'leaked' by Budd Hopkins. Whitley Strieber's alien abduction books, including *Communion*, followed a few years after. Prior to the publication of these books, the ubiquitous 'Gray aliens' had never been seen before. In fact, a review of the history of 'contact' cases shows that the type and variety and behavior of 'aliens' around the world are quite different across the board. But, along came Budd, followed by Whitley and his glaring alien on the cover, and suddenly the Grays were everywhere.
>
> In respect of Whitley and his Grays, allow me to emphasize one of Dolan's comments quoted above: *"By early 1969, teams within the CIA were running a number of bizarre experiments in mind control under the name 'Operation Often'. In addition to the normal assortment of chemists, biologists, and conventional scientists, the operation employed psychics and experts in demonology."* This, of course, brings us to the parallel event of that period of time: Satanic Ritual Abuse. SRA is the name given to the allegedly systematic abuse of children (and others) by Satanists.

As it happens, keeping our timeline in mind, it was in the mid- to late-1970s that the allegations of the existence of a *"well-organized intergenerational satanic cult whose members sexually molest, torture and murder children across the United States"*, began to emerge in America. There was a panic regarding SRA triggered by a fictional book called *Michelle Remembers*. The book was published as fact but has subsequently been shown by at least three independent investigators to have been a hoax. No hard evidence of Satanic Ritual Abuse in North America has ever been found, just as no hard evidence of abductions by Gray aliens has ever been found. Nevertheless, the allegations were widely publicized on radio and television talk shows, including Geraldo Rivera's show.

Religious fundamentalists promoted the hysteria and, just as during the Inquisitions, endless self-proclaimed 'moral entrepreneurs' both fed the fires of prosecution and earned a good living from it. Most of the early accusations of SRA were aimed at working-class people with limited resources, and with a few exceptions, the media and other groups that are ordinarily skeptical either remained silent or joined in the feeding frenzy of accusations. The few professionals who spoke out against the hysteria were systematically attacked and discredited by government agencies and private organizations …

Now, let's go back and think about our timeline. As it happens, *Michelle Remembers* was published in 1980, co-written by Michelle Smith and Lawrence Pazder, M.D. Budd Hopkins finished *Missing Time* in December of 1980, with an 'afterword' by Aphrodite Clamar, Ph.D. It's looking pretty 'coincidental' from where I sit.

Of particular note in the above passage is the fact that *"with a few exceptions, the media and other groups that are ordinarily skeptical either remained silent or joined in the feeding frenzy of accusations. The few professionals who spoke out against the hysteria were systematically attacked and discredited by government agencies and private organizations."* Keep this in mind as we review what the media was – at the same time – publishing about science.

The *Time* magazine article mentioned above gave more background on the Reagan/AAAS affair noting that Reagan had said that if evolution is taught in public schools, *"the biblical story of creation should also be taught."* Many educators interviewed for the article were shocked at the upsurge in biblical fundamentalism (why should they be shocked with the promotion of SRA and alien abduction going on in the background?) and its effect on society, mainly education. *Time*'s writers emphasized that many creationists actually believe that the world was created in 6 literal days only about 5 to 10 thousand years ago.

The problem here is: why is there even a problem? As Fix notes, of all the straw men that science has been tasked with knocking over, the belief in the recent creation of the earth in any context should be among the easiest, especially if the evidence for evolution is as overwhelming as they claim.

Part of the problem was the AAAS itself and its senior members and spokesmen. For some reason that we can only guess, considering what I have included about COINTELPRO above, these individuals were being set up as spokesmen for the solution to – or antithesis of – the burgeoning spiritual crisis that was taking place in society. The *Time* magazine article quoted paleontologist Stephen Jay Gould as saying: *"That evolution occurred is a fact. People evolved from ape ancestors even though we can argue about how it happened. Scientists are debating mechanism, not fact."* Considering that ordinary people were being scared out of their wits by alleged spiritual attackers in the form of Satanists and satanic aliens, on prime time TV no less, for someone like Gould to come along and say what he said only gives the average person the impression that he – and all his scientific fellows – are personal representatives of Satan himself

The fact is, the evolution issue is not quite as secure and settled as the AAAS and some of its senior members claim. Internationally respected paleontologist Bjorn Keurten, also an evolutionist, thinks that the human line has been separate from that of the apes for more than 35 million years and that it is more likely that apes descended from early man-types.

The troubling thing about this is the evident propagandistic nature of it. That is to say, Gould is dogmatic and ignores the fact that some very eminent scientists do *not* subscribe to the party line of the AAAS. Gould does not even attempt to make the elementary distinction between an opinion and an unequivocal assertion that may not be true. And so, lines are being drawn in the sand for some reason that we hope to understand as we proceed.

There is more evidence that this was a very definite propaganda program. The April 1981 issue of *Science Digest* carried an article by science journalist, Boyce Rensberger , wherein he stated:

> Except for those who believe in miracles of special creation (events that, by definition, can neither be proved nor disproved), no one doubts that our heritage can be traced back nearly four million years to little creatures that, as adults, stood only about as tall as a five-year-old today. ('Ancestors: A Family Album')

Rensberger's statement that *"no one doubts"* is extraordinary. (I would like to mention that 'the Big Bang' is definitely one of those 'miracles of special creation' that can neither be proved nor disproved.) The fact is that even among leading paleontologists, there are many, many doubts about the human lineage. Certainly Rensberger, as a science journalist, knew that. Certainly many educated laypersons knew it. And then Fix asks the question: *"Why em-*

Creationists claims are as dubious as those of Darwinists.

ploy such language? It seems almost calculated to discredit the subject."

Fix recounts a veritable flood of published material from the '70s and '80s that seemed to be more of a promotion than an even-handed assessment of paleoanthropology. Late 1980 – the same time as the media promotion of SRA and alien abduction – was also the time of the publication of Carl Sagan's *Cosmos*. The fact is, much of what Sagan wrote was superficial and glib, but more damning is the fact that he presented Darwin's ideas as though nothing had been thought about or reconsidered since then, though most certainly that is not the case. Sagan writes: *"The secrets of evolution are death and time – the deaths of enormous numbers of life-forms that were imperfectly adapted to the environment; and time for a long succession of small mutations that were by accident adaptive, time for the slow accumulation of patterns of favorable mutations."*

Sagan emphasized the term *"by accident"* by italicizing it. He was making it clear that accident is the fundamental factor in the emergence of life forms including, and especially, the very emergence of life itself in the vastness of the cosmos. Fix writes:

> ... Sagan invokes accidents the way others invoke God. His position implies that not only man but the entire universe is, in the final analysis, merely the result of a series of billions of accidents over billions of years.
>
> ... Scientifically, this position is no more capable of proof or disproof than miracles of special creation, and all the laws of probability are dead against it. Philosophically, even a child can see there is too much order and pattern in nature for this to be credible. It used to be said that God geometrizes. Do accidents geometrize? In Sagan's accidental cosmos, consciousness becomes an excretion of matter ...
>
> ... After years of reading scientific and quasi-scientific literature, it was suddenly clearer to me than ever why society has developed the cultural stereotype of the mad scientist. ...
>
> ... Sagan insists on playing the high priest of materialism and does it badly. If this opinion about accidents is perceived as the reigning scientific wisdom, it is

not surprising that many people have forsaken the counsels of science entirely ... the scientists themselves are driving them away in the first place with vacuous absurdities.

When Sagan excludes even the possibility that a spiritual dimension has any place in his cosmos – not even at the unknown mysterious moment when life began – he makes accidental evolution the explanation for everything. Presented in this way, evolution does, indeed, look like an inverted religion, a conceptual golden calf ... (Fix 1984, xxiii–xiv)

If this is seen in the context of what I mentioned above – the media promotion of SRA and alien abduction – things begin to look very strange indeed. With the people being worked up about spiritual matters on one side, and the scientific establishment being set up on the other side to appear as agents of Hell and Damnation, and the media behaving like demons with pitchforks jabbing first one side and then the other, it's no wonder that the average person in the U.S. has fled in terror back to the 'faith of our fathers'. In short, the exchange between Reagan and the AAAS was just the tip of the iceberg of much more serious problems in science where some force within it is working to discredit it rather than enhance it. Fix writes:

Scientists almost invariably overstate the case for evolution, and creationists have as much access to the critical literature as anyone else. But when it comes to defending their own theory, creationists do not have a strong case in terms of biblical scholarship or for a recent creation of the earth. And yet the creationists are stronger in the 1980s than they have been in many decades, and their influence appears to be still on the increase. (Fix 1984, 233)

Back in 1984 when this book, *The Bone Peddlers*, was published, Fix was noticing the upsurge of a phenomenon that has become a frightening reality in our own day – witness George Bush and his use of fundamentalists to justify his pre-emptive war policy, not to mention his shredding of the Constitution and institution of what amounts to martial law in the U.S. Sarah Palin and her Dominionist background is another frightening feature of this upsurge. Fix continues to pursue the problem and actually identifies the root of it:

The most peculiar aspect of the debate is that the creationists are sustained by events completely outside the question of evolution. The Bible of course does not end with Genesis. Scattered throughout the Old and New Testaments are a number of prophecies indicating that the nation of Israel would be reborn with Jerusalem as its capital and that there will eventually be a great, final, and dramatic last battle between Israel and an alliance of various nations, including one "from the uttermost parts of the north," (Ezekiel 39:2), a reference many see as pointing to Russia.

As everyone knows, Israel has in fact been reborn. As recently as 1980, the Israelis proclaimed Jerusalem as their eternal, indivisible capital. And almost every week the political realities in the Middle East can be seen falling ever more precisely into the pattern required for consummation of the ancient prophecies that are yet to be fulfilled.

This obvious and enormous fact hardly ever enters the argument, and yet the contemporary Middle East situation probably has more to do with the resurgence of fundamentalism than any other factor. (Fix 1984, 233–234)

The issue of evolution itself is not something that we want to get into here, but let me just quote another passage from William Fix, who writes:

When Kelly Segraves claimed in the California court that evolution was a secular religion, there was more to the charge than courtroom rhetoric. Indeed, he may have touched upon the central issue of the debate. In a detached, wider perspective it is almost impossible not to conclude that what we are witnessing is a conflict between competing systems of religious belief. (Fix 1984, 208)

And that is the crux of the matter: a totally materialist view of reality versus one that includes consciousness as something that can be non-material. The creationists have been deliberately provoked into reacting against the totally materialist view by the Jewish-controlled media, retreating into their 'God-did-it-in-6-days-and-it's-true-because-he-said-so' version. This suits the Zionist agenda perfectly.

What is troubling is the way the materialist evolutionists have been provoked into their untenable stance, which manifests equal determination to hold the line against any single acknowledgement of any process that is not totally random, accidental, and material. Their story is that the Big Bang was the explosion of a primal atom and all matter in the universe was in this incredibly dense atom. Everything that has happened since is just the result of random jostling of particles that, over billions of years, may form affinities by accident, and different forms of matter arise. Eventually, some of this matter jostles against some other bit of matter, some sort of electrical (or other) interaction takes place, and that is 'life'.

The Big Bang theory is creationism. Materialists believe that matter sprang suddenly into existence with nothing prior. That primal atom was there, and they make no attempt to explain it. That's as crazy as saying 'God was just there' and decided to create the universe.

Evolutionists argue: *"Creationists believe that the mind sprang suddenly into existence fully formed. In their view it is a product of divine creation. They are wrong: the mind has a long evolutionary history and can be explained without recourse to supernatural powers."* (Mithen 1999, 10) As you can see, archaeologist Steve Mithen is arguing from as false a premise as the 'God-did-it' gang. He has already made a big leap of assumption that when anyone speaks about 'mind' that they are speaking exclusively about a mind that is tied to a physical body.

The fact is, it is not that examples of natural selection cannot be found or observed, or that evolution has not been shown to have robust explanatory powers in some contexts; it is that *an equal, or greater, number of contradictory instances can also be cited.*

A close study of the matter indicates that the scientists at the forefront of research have actually gutted classical Darwinism, they just haven't – or aren't allowed to – go public with this news. It exists only in their scholarly papers and inner counsels. At the same time, the public is not able to distinguish between experts and popularizers and naively assume that sensational articles on the topic are based on demonstrated fact. More than one responsible scientist has voiced concern that the real facts about Darwinism and evolution are simply not reaching the public.

Proponents of Darwinism or neo-Darwinism insist that there are clear distinctions between science and religion. Indeed, there are obvious differences in the style and content of a laboratory experiment and a claim to divinely revealed knowledge. But if you look just a bit deeper and ask what is the belief structure of the person conducting the experiments, you may find, particularly in the U.S. (which is a problematical observation in itself), that the experimenter subscribes to the variety of evolutionary theory that is based upon an exclusively materialistic or mechanical hypothesis. Such an individual will distinguish science from religion by saying that science is concerned with knowledge of the proven and visible, while religion is concerned with mindless faith in the unprovable and invisible.

It goes without saying that a geologist identifying a stratum of fossils is on firmer ground than a theologian discussing the Trinity but here's the rub: it is of greater relevance to take natural selection as Huxley, Sagan, Gould, and others, employ it and ask: is natural selection really a proven fact based on demonstrated knowledge or is it an unproven hypothesis to which there are so many contraindications that belief in it is also, in the final analysis, only a matter of faith? Natural selection is no more visible than a deity.

Evolutionists are often found taunting creationists that their miracles of special creation can, by definition, be neither proved nor disproved. Yet the evolutionists arrive at similar propositions, especially when they exclude any possibility of something that guides and propels evolutionary processes. Karl Popper remarked of such theories in *Conjectures and Refutations* (1963) that *"A theory which is not refutable by any conceivable event is non-scientific."* The main difference between the believers in miracles of special creation and believers in accidental variations is that the former has God pulling the strings and the latter has only jostling atoms and molecules as its ultimate reality. Not much difference, eh?

It seems evident that evolution does function as a secular religion for many people. When they use the phrase 'no one doubts', they are implying that some ultimate revelation has been received that can only be understood by their high priests and devotees.

Alfred Russell Wallace, the co-founder of the theory of evolution, eventually came to the conclusion that natural selection could not account for man himself. He wrote that *"nature never over-endows a species beyond the demands of everyday existence."* This means that there is a major problem in accounting for many aspects of human beings – at least for some human beings. Stephen Jay Gould writes, *"The only honest alternative is to admit the strict continuity in kind between ourselves and chimpanzees. And what do we lose thereby? Only an antiquated concept of soul ... "* Here Gould is expressing the core of evolutionary materialism, *"the postulate that matter is the stuff of all existence and that all mental and spiritual phenomena are its by-products."* This is the pivot of the debate. What is more, as Zinoviev notes, this reduction of all mental and spiritual phenomena to 'by-products' of matter is no longer limited to biology and anthropology; it infects most of modern philosophy, the psychological and medical sciences, social systems, politics, and more. And this belief in evolution works to limit research in such a way as to confirm their basic 'postulate'.

G. G. Simpson of Harvard was one of the most thorough and extensive writers on evolution. As he put it in *Tempo and Mode in Evolution* (1944): *"... the progress of knowledge rigidly requires that no non-physical postulate ever be admitted in connection with the study of physical phenomena. ... the researcher who is seeking explanations must seek physical explanations only ... "*

The late Weston La Barre, professor of anthropology at Duke University, was consumed with ideological fervor against the 'enemy' and wrote that all religions other than evolution are maladaptive retreats from reality. When considering the Platonic philosophy which holds that ideas, forms, patterns, types and archetypes have an existence and reality of their own and would, therefore, seem to have an obvious relevance to evolution and the origins of species, he regularly compared Plato to Adolf Hitler. He neglected to mention – as such true believers do on both sides of the argument – that Hitler was a confirmed, even extreme, Darwinist, believing that man evolved from monkeys, a proposition that Plato would have considered absurd.

The writings of many great researchers, including Carl Jung and physicists and mathematicians, suggest that Plato was correct and that there are immaterial realities such as souls, archetypes, consciousness independent of physical brains, and more. The evidence for this is actually more considerable than the rags and tatters of evidence that are glued together to attempt to validate macroevolution. And, of course, this means that the advocates of materialistic Darwinism are the ones who are laboring under one of history's greatest delusions.

Quantum physics indicates that not only does 'matter' seem to dissolve into patterned vibrations at the most fundamental levels; it has become apparent that there is a structuring role played by consciousness. There is now much accumulated evidence that mind does exist separate from the physical brain and that the phenomena such as telepathy are not only demonstrable, but they conform to models of the universe with non-local causes.

In other words, the world has changed under the materialist evolutionist's feet and there is much more to our reality than the naive realism upon which neo-Darwinism is based. The fact that most contemporary evolutionists still cling to the old-fashioned, crude and mechanical theories in spite of the well-known developments in other scientific fields is more proof of the religious character of their beliefs.

And here we come to an interesting idea: the difficulty for both the believers in purely mechanistic evolution and the creationists is that any cosmology that is sufficiently explanatory of the phenomena we observe in our universe has deeper dynamics and implications. The evolutionists and creationists both do not seem to be capable of the truly abstract, subtle thinking required to parse these implications. It is as though both types are confined within a set of cognitive restrictions that drive their perceptions, experiences and priorities. When we collect the data on these types of individuals – and they are found in all classes and professions – we find a certain common factor that has been identified as the 'authoritarian personality type'. This type of individual is characterized by three attitudinal and behavioral clusters that are interrelated.

1) *Authoritarian submission* – a high degree of submissiveness to the authorities who are perceived to be established and legitimate in the society in

which one lives or the peer group with whom one is involved or works. Naturally, if many of one's peers are personality disordered – and there is a high probability of that being the case in any field where authority over others is to be had – then one is inculcated into submissiveness to psychopathological ideation.

2) *Authoritarian aggression* – a general aggressiveness directed against other people that are perceived to be targets according to the established authorities as defined in number 1.

3) *Conventionalism* – a high degree of adherence to the traditions and social norms that are perceived to be endorsed by society, or one's peer group, and its established authorities, and a belief that others in one's society should also be required to adhere to these norms. Once again, if the traditions and social norms are established by authorities with control agendas, everything is corrupted from the top down.

> According to research by [psychologist Robert] Altemeyer, authoritarians tend to exhibit cognitive errors and symptoms of faulty reasoning. Specifically, they are more likely to make incorrect inferences from evidence and to hold contradictory ideas that result from compartmentalized thinking. They are also more likely to uncritically accept insufficient evidence that supports their beliefs, and they are less likely to acknowledge their own limitations. (Wikipedia, 'Rightwing authoritarianism')

As we can see, this description applies equally to both sides of the 'debate', each rigidly submitting to their own chosen 'authorities', over-identifying with their own peer groups, rejecting conflicting evidence and attacking those who bring such evidence forward. Keep the authoritarian personality in mind, please, as we proceed. Remember that it is not necessarily identified by *what* is believed, but by *how* it is believed and promoted.

10. The Cycle of Ages

It's time to shift gears, but before we do, let me make my own position clear. I do not believe in 'God'. I do not believe that a god created this cosmos. I do believe, however, that pure, infinite potential known as Consciousness is the foundation of all that exists. I believe that we are 1) quantum; 2) chemical; 3) biological beings – in that order – and the quantum part of us is not material. My views broadly reject the premises on which both Darwinism and Creationism are based. However, I *do* embrace many of the ideas of evolution because they are empirically evident, but the basic theory of evolution as formulated by Darwin has been repeatedly shown to be impossible. I reject the theory of the Big Bang which is, in the final analysis, just pure Creationism under another name, another belief system. Both Creationism and the Big Bang theory posit that the universe is created, linear and finite. An eternal, infinite, *conscious universe* does not require a creator – it *is* what it *is*. Finally, I don't claim that this is the Truth; it is just where I am now after 45 years of research. There is certainly much more that needs to be discovered and understood. Now, let's return to our topic.

One of the problems with the view I have presented so far is that I have been working strictly off of mainstream archaeological and geological evidence. It's a useful view, of course, but I think we can figure out by now that it is not the whole picture. That is why I have not yet engaged the topic of whether or not a prior high civilization existed at the time of the Great Deluge. Modern, mainstream science denies that this is possible because, mainly, the theory of evolution tells us that earlier means primitive. (As mentioned, I don't want to divert to an analysis of the theory of evolution here. I'll just recommend a few very good scientific books on the topic: Bryant Shiller's *Origin of Life: The 5th Option*, Fix's *The Bone Peddlers* and *Shattering the Myth of Darwinism* by Richard Milton. The bibliographies in these books will take you to other excellent scientific works by mainstream scientists that will reveal the fact that there are still a few real scientists not under the political control of the AAAS and other corrupt pseudo-scientific organizations.)

The problem of deliberate falsification is, as I mentioned, a sociological process that deserves some attention. It's easy enough to say that everything we know – or think we know – is either false or twisted; what concerns us here is how this happened. What kind of society does that sort of thing and what are the processes by which it happens? Is there a way to reconcile the material brought forward by the revisionists with the facts presented out of context by the mainstream falsifiers as described by Zinoviev above? And most important of all: is there a relationship between what is happening in our world today and what may have happened in a previous high civilization that fell? Are we replaying 'Atlantis'?

The myth of a lost paradise, or Golden Age, is almost universal around the globe. The Golden Age was supposed to be a time when humanity was pure and good, and evil was unable to achieve dominance. The most well-known treatment of the theme is found in Greek mythology. The Greeks explained that life and civilizations run in grand cycles, rather like the seasons of the year. There was the Golden Age which, after a time, began to decline and gave way to the Silver Age, which declined to the Bronze Age, which declined into the Iron Age which, eventually, became Darkness. This is echoed in the Hindu or Vedic culture, with its model of cycling Yugas: The Satya Yuga (Golden Age), Treta Yuga (Silver Age), Dwapara Yuga (Bronze Age), and Kali Yuga (Iron Age). We are said to be living in the final days of the Iron/Dark Age, the Kali Yuga, at the end of which Brahma will awake from his dream and the whole cycle will begin again. The length of these ages varies from one system to another and have been interpreted and re-interpreted – much like biblical texts – to support various theories at various times.

The theme is represented as the Garden of Eden in the Judeo-Christian tradition, and that is where we find a most interesting twist: women are supposed to be the cause of the Fall. Indeed, most of the world's creation myths speak about some sort of ritual fault that occurred to bring about humankind's Fall from Grace and many of these myths attribute this fault to something having to do with sex.

In addition to the Fall from the Edenic state, there is also the Flood of Noah and the destruction of Sodom and Gomorrah, which is blamed on sexual excess. In the Sumerian myths, it is said that humans copulating all over the place kept the gods awake and they could get no rest, so they resolved to destroy the whole kit and caboodle.

Many authors of revisionist history write about Atlantis as though it were the Golden Age. However, there is a problem with positing the civilization of Atlantis as the exemplar of the Golden Age because, in Plato's exposition on Atlantis, he emphasizes the fact that it was an evil empire that instituted a world war (though no mention is made specifically of sexual excess or the depravity that had taken over Sodom and Gomorrah). But I have already suggested that there was a way of life prior to the Deluge that was rich, fulfilling, and stable – without warfare – for many thousands of years. So, if that was the case, why the Deluge?

My thought is that Atlantis was the end-point of a cycle and the Upper Paleolithic – before the great technological civilization of Atlantis came into being – was the Golden Age. Fully actualized Atlantis was the 'Iron Age/Dark Age' of that cycle and it was in the process of trying to institute a New World Order when the destruction came. If I am correct, then we may come to some approximation of the length of the cycle as well as some idea as to where we are on that cycle, and thus, of our future.

The solution to sorting all this out is, I think, the solution to the problem of reconciling mainstream archaeology with revisionist histories. Remember what Zinoviev wrote about the mainstream falsifiers? *"[T]heir forte is the ability to misrepresent historical events* **while giving correct temporal and spatial coordinates and representing individual facts veraciously and in full detail.** *The actual falsification is achieved via the* **selection of facts,** *their* **combination and interpretation,** *as well as the* **context of ideological conceptions,** *propagandist texts that they are immersed into ..."*

From this we may assume that the facts and locations given by mainstream paleontologists, archaeologists and geologists are accurate as far as they go: it is the selection – they leave out a lot – and the way they combine and interpret them and place them in an ideological framework (evolution), that is the mode of deception. Having something to work with, in the course of sorting all this out, we may also come to some idea of how it is that our civilization is so completely dominated by liars.

There is clearly a substantial body of evidence that a sort of 'superman' human being known as Cro-Magnon appeared suddenly in Europe and he was something that had never been seen before in the previous 200 thousand years (or more) of the existence of Neanderthal man. The evidence shows that this new man was the creator of cave paintings and other art that is astonishing in view of the alleged primitiveness of his lifestyle. The connection between the so-called primitive lifestyle and the paintings is pretty solid as far as I can see, time-wise. However, as noted, this was many thousands of years before the Deluge, so there was plenty of time for an advanced civilization to develop after this, and it is in the development of this technological civilization that I believe we find the 'ritual fault' that brought about its demise – and it may definitely be related to 'sex', though not in an ordinary sense.

The revisionist researchers have certainly assembled plenty of evidence for an ancient, high civilization. This evidence is totally ignored by mainstream science. They explain it away in the most ridiculous ways, and if they can't do that, they engage in *ad hominem* attacks. Most of this evidence for great cities appears to date well after the appearance of Cro-Magnon and his obviously 'primitive' lifestyle – though that assessment of his lifestyle as 'primitive' may not be correct. When we consider planetary cataclysm and its ability to scrub the surface of the earth clean, we have to understand that there's not going to be a lot of evidence to examine. That there exists any at all is a miracle. We have to work with what we have, but let me say that the revisionists have more solid evidence on their side than the evolutionist mainstream pseudo-scientists even if the latter claim the degrees and the methodology (which is admittedly good) and all the grant money.

I've given the whole 'cycle of ages' concept a lot of thought after reading everything I could on the topic and here's what occurs to me, though it is only a first approximation and is going to need some further work: There are many cycles within cycles that are experienced by life on this planet, by the planet, the solar system, the galaxy, and even the universe. The longest cycle that could relate to the earth as part of the solar system is its orbit around the galaxy which takes about 225 million years. I would call that an 'Aeon'. That is nine thousand 25,000-year precessional cycles. One third of 9000 precessional cycles is 7.5 million years.

There are a number of ways to deconstruct these numbers and the most satisfying is to reduce the Aeon period to 750 periods of 300,000 years each. Let's call these 300,000-year periods 'Epochs'. Next, 300,000 can be divided into 4 'seasons', or periods of 75,000 years each, which we can call 'Great Eras'. The same period is further divided into 12 periods of 25,000 years each, which is approximately a precessional cycle or 'Great Year'. A precessional cycle can be divided by four (the four Yugas/Ages giving a period of 6,250 years for each Yuga). See the first graphic below.

The precessional cycle can also be divided by 12, giving a period of 2,083.3333 years. These latter we refer to as astrological signs, or 'Earth Ages'. Three signs belong to a Yuga/Age. See the second graphic below.

You can take any of the segments referenced and depict it as its own 'great cycle' and further divide it down into smaller and smaller units. I have done this with the second graphic. In the first graphic, you do not see the Aeon because that is simply too long a cycle to be concerned with. The largest one represented is one of the 750 Epochs. But in the second graphic, I have reduced the time periods so that each astrological sign period of 2083.333 years is divided into four Earth Eras of 520.8333 years. I thought I would play with it a bit and make 2013 the end of the current Iron/Dark Age/Earth Era and began subtracting these Earth Era periods and, as you can see, the numbers turned out to be very interesting.

The first subtraction gives us 1492 which is when Columbus 'discovered' America. The second date, 971, is the year the Mosque Azhar was built in Cairo,

Egypt (970–972). It is believed that the Norse explorer, Leif Eriksson, was born in this year. What we do know is that, at the beginning of the ninth century, Celtic Britain lay in ruins. This suggests that there had been destruction of some sort prior to this time. Nennius compiled a history of his people that makes fascinating reading about this time, though it is often rejected as pure confabulation by 'experts'. Perhaps it wasn't?

The next date that comes up at a transition point is 450 AD. This was the time of Attila the Hun and the fall of the Roman Empire. The date 1633 BC is the approximate time of the eruption of Thera and, in my opinion, the time of Abraham and Sarai in Egypt, known there by the names Ay and Nefertiti. The point is that, even if it is only a first approximation, this appears to be a very interesting way to look at the cycles of time.

Getting back to our topic, there is a great deal of uncertainty about the dates of various types of early humans. What I want to say is that the Golden Age of legend was a Great Golden Era as depicted in the first graphic and that more than one civilization has arisen and fallen in the 300,000 years since the pres-

ent Epoch began. Perhaps the Great Deluge, the alleged destruction of Atlantis, estimated to be 12 to 13 thousand years ago was just the ending of an age within an age within an age?

Perhaps entire sub-cycles of ages as described in the theory of the Yugas played out between the time of the first appearance of Cro-Magnon in Europe and the end of the Pleistocene around 12,000 years ago? I don't think that we can consider the life of humanity either before or immediately after the Deluge to have been a Golden Age by any stretch of the imagination. But then, if we consider the Upper Paleolithic as a sort of Golden Age (and we will see that we have some reason to do so), and our current age, the Iron Age, moving toward Darkness and Destruction, then where does the Deluge fit into all of this? Well, half of a precessional cycle is 12,500 years and many experts working on this issue of cometary impacts and the Deluge say that this is when it occurred: 12,500 years ago, often abbreviated to 13kya. Why is this larger cycle I am proposing important? Let's consider the very big picture.

In their book, *The Sixth Extinction*, Richard Leakey and Roger Lewin describe the present as a time in which we are moving toward the extinction of at least 50% of all life on earth, possibly including humanity. They suggest that most of the extinction is occurring due to human overpopulation and overexploitation of the planetary resources. But there is more to this than what can be blamed on humanity as the research on cometary impacts clearly reveals. Our current extinction is, apparently, just part of a very large cycle. Richard Firestone *et al.* write in *The Cycle of Cosmic Catastrophes*:

> When people use the term *extinction*, they mean that many living species vanished. This is just part of the equation, however. Another side is that some species survived. In all past major extinctions, with ecosystems out of balance, many of the surviving species experienced explosive growth. This is what happened 13,000 years ago, when an unusual mix of conditions created favorable conditions for the human species.
>
> First, there were new genes. Spurred by genetic mutations that produced a burst of creativity and technological resourcefulness, humans became even more efficient predators. The Event killed off many competing predators, and skilled hunters decimated many surviving ones, ensuring that more people survived. The same resourcefulness made humans better at finding food in every way, and consequently human populations expanded rapidly.
>
> Second, a benign climate, due partly to the greenhouse effect of the impacts, fostered larger human populations. The warmer climate, coupled with humans' newfound resourcefulness, fostered the invention of agriculture, freeing humans from a nomadic lifestyle. The development of better housing, clothing, and weapons all allowed human populations to increase.
>
> Third, increasing populations led more people to live together in villages and towns, where the division of labor allowed a larger pool of skilled talent to develop. This fueled an almost constant technological boom in many fields, producing, among other things, pottery making, metalworking, and writing.
>
> All that may seem positive, except that the burgeoning population, initially fostered by the extinction Event, contained the seeds of many of our current troubles. When overpopulation occurs in any species – whether it is rabbits, locusts, lemmings, or people – a host of problems comes along with it, including epidemics, starvation, extreme aggression, ecosystem destruction, and scarce resources, every one of which is a major pressing problem in our society today.
>
> ...
>
> An extinction sequence comprises the following stages:
> - A major catastrophe leads to the disappearance of some species.
> - These disappearances lead to the overpopulation of some surviving species.
> - Overpopulation leads to devastating depopulation.

This equation has held true for every past extinction event. In the current sequence, we have passed through the first two stages as a species but not the last stage, depopulation. (Firestone *et al.* 2006, 186–187)

There is more to this than Firestone *et al.* suppose, I think. But it is important to us to realize that what happened 13,000 years ago was just the one of many extinction events. If this unbelievably violent cataclysm – in which almost all life on earth was destroyed – came as a global empire was seeking domination over others, we may need to ask the question: does our living planet, its companions in the solar system, and its parent sun, manifest some sort of consciousness and do they, between them, have the resources to deal with humanity – or any other species – whenever things get 'too hot', so to say? Was Atlantis such an example? Did other destroyed and buried civilizations whose names were completely forgotten for hundreds, if not thousands, of years, become victims of their own hubris in like manner?

Getting back to the possibility of the existence of an advanced civilization called Atlantis, this has been argued for a very long time and many theories have been proposed. I'm not going to examine that here; I'm just going to highlight the fact that this is still an ongoing scholarly discourse. There is an interesting paper by Emilio Spedicato of the University of Bergamo entitled 'Galactic Encounters, Apollo Objects and Atlantis: A Catastrophical Scenario for Discontinuities in Human History'. It was first published in 1985, revised in 1990, and again in 1997. He writes in his summary:

Recent findings about interactions of the Earth with extraterrestrial bodies, particularly comets and Apollo-like objects, are reviewed, with special attention to climatological effects. We discuss the hypothesis that the last glaciation was started by a collision over a continent and was terminated by a collision over an ocean. We propose that during the glaciation sufficiently good climactic conditions in the lower latitudes made possible for mankind to develop a high level of civilization. The Platonic story of Atlantis in interpreted as an essentially correct description of a political power active in the final period of the last glaciation.

A political power active in the final period of the last glaciation with designs on taking over the world doesn't sound much like our peaceful Cro-Magnon peoples who painted the caves and produced stonework to rival the most talented sculptors of our day and who maintained a homogeneous lifestyle that was so satisfying that it lasted many thousands of years according to the evidence. It certainly sounds like something happened between the time of the cave paintings and the Deluge. It could be the same thing that is happening in our own culture: a creeping corruption that slowly, but surely, poisons everything and leads an entire civilization to destruction. So, it may behoove us to have some idea of what this might have been, how it may have happened. Our discussion of the corruption of science above may have a great deal to do with this.

Everyone knows that there is something terribly wrong with our world, but there are hundreds, if not thousands, of explanations for why this is the case. Some of them are right or partially right; some of them are just plain wrong and designed to keep people ignorant. From my own point of view, ascertaining the truth about our history – as close as we can get to it – is a very important task in the process of figuring out what is wrong. You can't figure out how to break out of something unless you understand the nature of it thoroughly – you kind of have to back-engineer things, which is where studying history and the natural sciences comes in. Plus, you have to have evidence sufficient to convince others that any plan you might have is based on fact – primary reality – and won't make things worse. A social physician has the responsibility to 'do no harm' and that means don't treat anything you don't understand.

I have theorized previously that the cometary destruction associated with the Deluge may have been responsible for mutations in the human population and these mutations led to a variety of psychopathologies that have spread gradually in the population, incrementally corrupting humanity and life on earth, until we are where we are today. That may have been what happened to Atlantis via an even earlier cataclysmic interaction, and I'm going to explain why I think so. What Atlantis ended up with, according to Plato, and what we have today, was and is a corrupt global system that seeks domination of the entire world. For Atlantis, in the midst of, or immediately after, fighting a war of domination, the Deluge came destroying nearly all life on earth in a single day and night. Rather sobering if you think about it.

> But as the days of Noë were, so shall also the coming of the Son of man be. For as in the days that were before the flood they were eating and drinking, marrying and giving in marriage, until the day that Noë entered into the ark, and knew not until the flood came, and took them all away … (Matthew 24:37–38)

So, something corrupted the world of Cro-Magnon man long before this time and whatever it was, it led peaceful hunter-gatherer types to create a complex technological civilization that then took the peoples of the world into an abyss; and this corruption may very well have survived the Deluge and still be present in our own society. It is a system that posits pure materialism as its foundation, and excludes entirely the potentials of consciousness as a factor in human dynamics (except as a by-product or 'excretion' of matter). And when I refer to materialists, I actually include the creationists because their position is actually as entirely materialism-based as the evolutionists. Let's look at a few facts that may reveal to us clues to the origins of this mode of thinking/belief. Paul Mellars writes in *The Neanderthal Legacy*:

> Perhaps the most intriguing and enigmatic aspects of the Middle Paleolithic period is how and why it came to an end, after a period of around 200,000 years of remarkable stability.
>
> From the preceding chapters it has emerged that while there were significant shifts in the precise morphology and technology of stone tool production, subsistence patterns, site distributions etc. at different stages of the Middle Paleolithic, very few if any of these seem to reflect any radical reorganization or restructuring of technological, economic or social patterns. Most of the documented changes appear to be more cyclical than directional in character ... none of these changes at present suggests more than a reshuffling of basic cultural and behavioural patterns which, in one form or another, can be traced back into the time range of the penultimate glaciation ...
>
> The dramatic break in this pattern of behavioural stability occurs at the time of the classic Middle-to-Upper Paleolithic transition, dated in most regions of Europe at around 35–40,000 bp. (Mellars 1996, 392)

And then he asks the most important question in all of archaeology and paleontology: What is the precise character of the behavioral change and to what extent was this due to a major dispersal of new human populations? The corollary question is, of course: why should we encounter this particular combination of biological and behavioral change at this specific point in the archaeological sequence? It was, after all, a time when large parts of Europe were still in the grip of an Ice Age. (Or, so we are told, though Allan & Delair make a convincing case for a different interpretation of ice ages in their book *Cataclysm*.) We note that the Spedicato paper cited above proposes that the last glaciation began with a cometary collision or explosion over land. Perhaps we find here a clue to the sudden appearance of Cro-Magnon man?

In a paper by Rhawn Joseph and Chandra Wickramasinghe entitled 'Comets and Contagion: Evolution and Diseases From Space' (*Journal of Cosmology*, 7, 2010, 1750–1770), the authors write in their conclusions:

> Correlation is not causation and thus no firm conclusions can be drawn despite the wealth of evidence suggesting a link between comets and diseases from space. Nevertheless, comets are an ideal vehicle for sustaining and transporting a variety of microbes, including viruses, from planet to planet and even from solar system to solar system. In consequence, when these organisms are deposited on a

223

world already thriving with life, genes may be exchanged, the evolution of new species may ensue, or conversely contagion may be unleashed, and disease, death, and plague may spread throughout the land.

Let us speculate that the genes that produced Cro-Magnon man may have been brought to earth as the result of a cometary impact. The simplest version of this panspermia theory is that proposed by Sir Fred Hoyle and Chandra Wickramasinghe who suggest that life forms continue to enter the earth's atmosphere, and may be responsible for epidemic outbreaks, new diseases, and the genetic novelty necessary for macroevolution. The mechanisms proposed for interstellar panspermia may include radiation pressure and lithopanspermia (microorganisms in rocks), and deliberate, directed panspermia from space to seed earth. Interplanetary transfer of material is well documented, as evidenced by meteorites of Martian origin found on earth.

There are several other things that tie into this. I would suggest that the reader might want to have a look at *Planet-X, Comets & Earth Changes* by James M. McCanney, which I highly recommend, and *Cataclysm* by D.S. Allan and J.B. Delair. McCanney's work is concerned with the Plasma Discharge Comet Model. These ideas are about electrical charges of comets and how they interact with the sun and the planets of our solar system. What is important here is what McCanney remarks about the planet Mars. He writes:

Mars is observed to have had oceans in the past and these are gone and the atmosphere that held the water and oceans and rivers in place is also gone. Why would a planet nearly the size of Earth with vast oceans and a substantial atmosphere holding it in place all of a sudden have it removed? ... The reality is that if this had happened millions of years ago, the vast dust storms that ravage Mars every year would have erased or eroded the geological features long ago ... In reality, Mars could not have lost its atmosphere more than a few thousand years ago, and there are historical records that the ancients saw a large comet pass by Mars and lift off the oceans and atmosphere in a single night ...

The ancients are said to have witnessed this and described it as a giant snake-like extension coming out of the huge comet that eventually became the planet Venus.

If a large comet nucleus comes near enough to a planet to interact electrically, that is, the comet establishes an electrical connection with the planet, then what is called the 'surface gravity' of the object is allowed to take over. If this happens the object with the larger surface gravity will suck at the volatiles (air, water, etc.) and if this process is allowed to work long enough, the object with the larger surface gravity will remove most of all of the atmosphere and oceans of the smaller object. One object is stripped of its cloak and the other receives a significant pollution event. It is evident that Mars recently underwent this … This also proves that the object was at least larger than the planet Mars – as large as the planet Venus or larger. …

Earth has in fact been involved in similar events but so far we have managed to 'win' the tug of war … receiving what amounts to vast pollution events (the rain of brimstone and burning naptha or hydrocarbons as well as water flooding). It is clear from this also that the world's oil did not come from millions of years of decaying fern forests (no chemist has ever developed a process to convert decayed matter into oil).

The oil fell onto the Earth during such a pollution event when the Earth performed celestial battle with a large cometary intruder. The great earth-wide flood was also the result of the water in the comet tail falling onto earth during the same or similar encounters … Most scientists do not realize that comets can become very large and that the volume of water, oil, etc. that is produced in the tail of a very large comet would dwarf the atmosphere of planet Earth. When the oil in the Mid-East was first used in quantity it was laying in pools on the ground. When this was used up, they dug into the sand and now we are drilling into oceans of oil that seeped into the sand after falling from above during the many comet encounters that Earth has seen.

Another aspect of the arrival of new objects from outside the solar system is that they many times arrive in multiples. This is because whatever the source of these wandering objects from beyond, they were formed or fragmented at the same time …

Since comets come in multiples … could it be possible that the massive comet Hale-Bopp was a precursor or companion …?" (McCanney 2003)

The important thing about McCanney's theory (and you have to read the whole book to fully appreciate it and understand why the corrupt scientific community sought to silence him) is the part about electrically charged bodies being able to strip a planet and then, when encountering another larger planet, the second planet can acquire all this material. This may definitely include organic material, possibly even atomized body parts of formerly living creatures containing bacteria, viruses, and more. This could explain the sudden appearance of Cro-Magnon man: a virus carrying a significant amount of DNA from a former human host on another planet could have infected a population of early hominids on earth with the result being instant modern man.

Of course, where the human types lived that said DNA came from is another question. Mars? Possible. This is where the Allan & Delair book, *Cataclysm*, comes in. This is probably one of the most excellent compilations of scientific data – sourced and referenced – I've ever encountered on the topic of our earth's cataclysmic past. As they note in their introduction: *"Some readers may recoil at the abundance of references and might even regard them as irksome or intrusive. In a work of this genre precise documentation is essential ..."*

Allan & Delair bring serious questions to bear on the mainstream interpretation of our reality and history and do it armed to the teeth with science. The case they make for a Golden Age world prior to the Deluge is compelling and quite unique. Wielding hard data from literally every field of science, they demonstrate that hundreds of thousands of years of ice ages may be a myth created to explain many anomalous findings on earth that uniformitarian science had no other way to explain. This data strongly suggests a completely different planet prior to a worldwide cataclysm that they say occurred in 9500 BCE, but the latest research puts the most recent major event back at least another thousand years. They refer to it as the 'Phaeton Disaster'. They also bring up the issue of the 'Fifth Planet' that is posited to have existed where the asteroid belt now lies. They propose that the Phaeton Disaster included the destruction of this Fifth Planet and the celestial calamity – which affected the entire solar system – left debris everywhere. Here, again, they posit that this occurred at the same time as the Deluge we have related to the destruction of Atlantis, 13,000 years ago, though they suggest a different date. I think that there may be some conflation of a number of separate events here. As noted, Spedicato proposes an event that initiated an Ice Age and another event that ended it. Other sources demonstrate compellingly the evidence of other great dyings on earth, so we don't have to confine ourselves to just one or two episodes.

In any event, Allan & Delair suggest that Phaeton was spawned in an astronomically-near supernova explosion and that it was a portion of exploded astral matter that came careening into our solar system, instigating 'war in heaven', destroying the Fifth Planet, taking its resources, engaging with the earth and dumping that water and oil and so forth on our planet while changing its axial orientation, completely reorganizing the arrangement of land and water (obviously, if we acquired more water, there was more land prior to that time and their map showing how it must have looked, based on scientific finds, is fascinating), then Phaeton plunged into the sun never to bother anyone again. While I really like nearly everything they have proposed in their book (and have supported with references), I'm not too sure about this idea. It's possible, of course, but Victor Clube and Bill Napier make a compelling case with their 'giant comet' that cycled around numerous times, and this ties in very well with what McCanney proposes (echoing Velikovsky): that the giant comet was Venus. The giant comet that breaks up into numerous other large comets described by Clube & Napier could then be chunks of the Fifth Planet running amok, and still there, for the most part, threatening life on earth. Of course, this again includes numerous events going back much further than 9,500 BCE. This is where the application of a lot of scientific minds would come in handy if they were allowed to work openly on such problems.

In any event, the important thing about their book is not only the detailed collecting of data that supports their interpretation of what the earth must have been like prior to at least one Deluge, but also their data (weaker because it is mythological, but still well done) about the Fifth Planet being involved in this disaster along with Mars and Venus. Their description of the antediluvian world is simply marvelous: a real Golden Age could have existed on such a world.

Now, just suppose that there were a Fifth Planet, where the asteroid belt now exists, in our solar system as so many have theorized. And just suppose this planet did have an advanced technological civilization of some sort. Suppose further that they were aware of the possible imminent destruction of their planet and a) launched ships to earth carrying a small number of refugees from space or, if there was too little time or their capacities weren't up to so large a project, b) launched engineered viruses toward earth that contained their own DNA. The other alternative is something like the scenario with Mars as described by McCanney: perhaps the intruder sucked off a great deal of material from said Fifth Planet and earth captured it in a subsequent interaction (along with material from Mars) and then dumped it on earth. This could have included a lot of organic matter, including viruses that had incorporated large segments of the DNA of the advanced beings on that planet which then infected a population of earth hominids, causing macroevolution in just a few generations. Such viruses as might be carried by space rocks could also be the explanation of the disappearance of Neanderthal. Just think 'Black Death'. I think the answers may be found in some combination of these various factors.

Each of these books contributes a great deal to the solution of the origins of mankind on earth, I think, as well as dealing with many of the problems of archaeology, paleontology, geology, and so forth. But there is more that we need to think about.

We know that the earliest known remains of Cro-Magnon-like humans are radiometrically dated to 35,000 years ago, though there are important finds indicating that modern human types may have existed on earth millions of years ago. Naturally, these finds are suppressed by the evolutionists (recall their missionary zeal to spread their religion). I have a problem with the date of the appearance of Cro-Magnon and here's why: In my book, *The Secret History of the World*, I cite the research of Richard Firestone and William Topping which demonstrates that there are a number of events, including cometary or asteroid impacts, not to mention a host of less dramatic things, that can –

and have – 'reset' the carbon dating clock (as well as problems with other forms of dating). Because we are fairly certain that there have been numerous such events, we have to assume that things that are radiometrically dated are probably much older, perhaps by a factor of two. So, I'm going to go with that for this hypothesis. (See my book *The Secret History of the World* for a detailed discussion of this problem.)

Let's suppose that the planet Venus was the comet in question that entered the solar system as McCanney proposes, supporting Velikovsky, though the 'birth of Venus' from Jupiter is rejected. Suppose this is the same giant comet of Clube & Napier only it entered the solar system much earlier than they suggest. Then, suppose that this giant comet did the things that Allan & Delair propose in their scenario and it did so, say, 70 to 75,000 or even 80,000 years ago. The reason I suggest this date is two-fold: 1) considering the dating problem revealed by Firestone and Topping, I think it is safe enough to think that most dates that paleontologists and archaeologists work with are off by a factor of two; 2) it fits neatly into the Epoch I have proposed above. If we are looking at a species extinction cycle, it must be a very long period and the 'Winter' season of this 300,000-year cycle consisting of 12 precessional cycles, began 75,000 years ago. Certainly, we could be anywhere along any proposed cycle; this is just the one I am proposing right now, speculatively, because a few things fit, including the paleontological findings about Neanderthal. The average time of his full manifestation probably falls right around 300,000 years ago, taking dating issues into account. (By 130,000 years ago, complete Neanderthal characteristics had appeared according to the evolutionists, whom we realize we cannot trust.)

There may or may not have been an advanced civilization on earth prior to that time during a previous Epoch. Considering what happens to the planet during cataclysms (read Allan & Delair for graphic descriptions) it's not likely that much, if anything, would survive, so we can't speculate much about that.

Going in another direction, I want to note again that paleontological finds suggest that modern-type humans did exist on earth at that time – and even earlier – though they were different from Cro-Magnon man. Anatomically modern human fossils – actually, they are *almost modern* – date back as far as 195,000 years. (We may consider the dating problem here, as well.) Research 'Omo remains', '*Homo sapiens idaltu*', and 'Skhul and Qafzeh hominids'. You'll find that they are all 'almost' modern.

So, this large chunk of rock comes into the solar system and becomes a comet due to its interaction with the solar capacitor according to McCanney's theory. It has the famed interaction with the Fifth Planet – Tiamat as recorded in the Sumerian stories – splitting it apart. A lot of it is left in smaller chunks and becomes the asteroid belt between Mars and Jupiter, another big chunk goes tearing off on its own as a 'giant comet' that may have disappeared for a while, but then was pulled back into ever smaller orbits until it broke up in the heavens and bombarded the earth 13,000 years ago. The same chunk gave birth to other comets and meteor streams as described by Clube & Napier, in earth-crossing orbits, and we have a date with some of the larger chunks in the not-too-distant future. That is, unless there is another player, which we will come to shortly.

Now, 'panspermia', as the DNA-transported-by-comets-seeding-life-on-earth theory is called, may get us off the hook as far as human evolution on earth is concerned, but it does not get us off the hook when considering where that DNA came from originally and how the individuals who carried it evolved, if the arguments against evolution that the panspermia scientists employ apply everywhere. Obviously, they are under the same constraints. On the other hand, that may be a way out in a different direction: DNA could be a pure manifestation of consciousness, a sort of first-level physicality, the interface between the material and non-material worlds. Pure information might be able to geometrize itself in the form of DNA and, *voilà!*, the building blocks of life that are complex and capable of inducting consciousness itself into matter come into being in an instant – sort of a mini-Big Bang, with consciousness present to guide the 'explosion'. Astronomer Sir Fred Hoyle wrote in a little book entitled *The Origin of the Universe and The Origin of Religion*:

> More than a century ago Alfred Russel Wallace noticed that the higher qualities of Man are acausal, like the Universe itself. Where human qualities have been honed by evolution and natural selection there is very little difference between one individual and another. Given equivalent opportunities for training, healthy human males of age 20 will hardly differ in their abilities to run at pace by more than 10 percent between the Olympic runner and the average. But for the higher qualities it is very much otherwise. From enquiries among teachers of art, Wallace estimated that for every child who draws instinctively and correctly there are a hundred that don't. The proportions are much the same in music and mathematics. And for those who are outstanding in these fields the proportions are more like one in a million. Having made this point Wallace then made the striking argument that, while the abilities with small spread, like running, would have been important to the survival of primitive man, the higher qualities had no survival value at all. Over a span of 12 years spent in the Amazon and in the forests of the East Indies, Wallace … lived with primitive tribesmen …What he said was that in his experience he never saw a situation in which an aptitude for mathematics would have been of help to primitive tribes. So little numerate were they that, in 12 years, he saw only a few who could count as far as 10. His conclusion was that the higher qualities, the qualities with large variability from individual to individual,

had not been derived from natural selection. Abilities derived from natural selection have small spread. Abilities not derived from natural selection have wide spreads.

Wallace was writing only a decade after Maxwell's work on electromagnetic fields, which even in physics were still seen as mysterious. It therefore seemed reasonable to speculate that perhaps there was some universal field which acted on matter so as to produce intelligence once evolution had advanced to a suitable stage. This is an idea which still resurfaces from time to time. But today it comes so close to offending areas of certain knowledge that for me, at any rate, it has no plausibility. **I think the higher qualities must be of genetic origin, the same as the rest. The mystery is that we have to be endowed with the relevant genes in advance of them being useful. The time order of events is inverted from what we would normally expect it to be, a concept that is of course gall and wormwood to respectable opinion ...**

In recent years I have managed to earn the dislike of respectable societies by saying that natural history was nearer to the truth in the first half of the 19th century than in the second half. (Hoyle 1993, 5–7)

In any event, we know that Cro-Magnon shows up in the archaeological record rather suddenly. We can't derive anything from that about how he may have arrived – bodily or as an infection, so to say. What we do know is that this was the most stunning event in human history. I'm not even going to quote the numerous experts who write about this event because everything they write reveals their helplessness to explain it. Many of them have the grace to admit it.

In any event, getting back to the Golden Age, after Cro-Magnon man arrived in Europe, the region apparently achieved a sort of nirvana civilization that was apparently peaceful and stable for over 25,000 years. But what happened then? If they were satisfied with their way of life, what happened to change it? If they went on to develop a technological civilization (or some of them did) and were responsible for the many incredible ruins of cities found all over the globe, what drove them to do that? And did only some of them do it in only some places? Were these the Neanderthal-human hybrids mythecized as Cain, Tubal-Cain, etc.? And then, if they developed this civilization as the myths and legends – and many global remains – tell us, what led them to warfare, to attempts to dominate the planet? And did this, in itself, have anything to do with the cataclysm that almost totally erased them from the face of the earth?

11. The Cro-Magnon Mystery, Neanderthal Man and Psychopathy

The appearance of Cro-Magnon man brought various technological and behavioral innovations, an explosion of symbolic artifacts, varied and sophisticated representational art, and more. All of these changes are associated clearly and unequivocally with this first appearance of fully modern human beings. The 'culture' of the Cro-Magnon peoples is called 'Aurignacian'. There was a striking uniformity of this Aurignacian technology over a vast area of Europe and the Middle East that appeared without any convincing origins or antecedents. There can be no doubt whatsoever that the appearance of the Aurignacian reflects *an intrusion of an essentially new human population*. And for a while, they lived in and amongst Neanderthal man.

The only way to fully know how extraordinary the new type of human being was, is to study carefully the old types, and that is what Paul Mellars' book *The Neanderthal Legacy* does so well. You will be left in no doubt whatsoever that Neanderthals are not our ancestors, though, indeed, there may have been some extremely limited genetic mixing as some recent DNA studies show. These studies suggest that this mixing was as long ago as 400,000 years and does not seem to have occurred in Africa, but rather in nearly all other areas of the planet except Africa. So, that's a problem to be dealt with.

A careful consideration of the characteristics of Neanderthals – what little can be discerned, but mainly their lack of creativity over 200,000 years or more of existence – along with the very small percentage of DNA mixing, may actually give us clues as to what may have happened that modern science is trying so hard to cover up with their evolutionary theories: genetic human pathologies including those that infect the very individuals who have taken over control of our civilization and corrupted it with their mechanical, materialistic world view.

It may very well be that certain personality pathologies among modern humans, such as the authoritarian personality and psychopathy, are due to the presence of some Neanderthal DNA that has recombined in particular ways. Here is why I suggest this: if you study all the available material on Neanderthals, you come to the idea that their basic natures – in the emotional/spiritual sense – are that of the psychopath. The following quotes from Robert Hare's book, *Without Conscience*, should make the comparison clear:

> Psychopaths are social predators who charm, manipulate, and ruthlessly plow their way through life, leaving a broad trail of broken hearts, shattered expectations, and empty wallets. Completely lacking in conscience and in feelings for others, the selfishly take what they want and do as they please, violating social norms and expectations without the slightest sense of guilt or regret. Their bewildered victims desperately ask, 'Who are these people?' 'What makes them they way they are?' 'How can we protect ourselves?'…
>
> Psychopaths have what it takes to defraud and bilk others: They are fast-talking, charming, self-assured, at ease in social situations, cool under pressure, unfazed by the possibility of being found out, and totally ruthless…
>
> Psychopaths are generally well satisfied with themselves and with their inner landscape, bleak as it may seem to outside observers. They see nothing wrong with themselves, experience little personal distress, and find their behavior rational, rewarding, and satisfying; they never look back with regret or forward with concern. They perceive themselves as superior beings in a hostile, dog-eat-dog world in which others are competitors for power and resources. Psychopaths feel it is legitimate to manipulate and deceive others in order to obtain their "rights," and their social interactions are planned to outmaneuver the malevolence they see in others. (Hare 1999, xi, 121, 195)

Depiction of Neanderthal life in the Krapina cave in Croatia.

The normal world of human impulses and reactions (e.g., emotional bonding, pro-social responses), concepts, feelings, and values strike essential psychopaths as incomprehensible and with no obvious justification. Ted Bundy called guilt *"an illusion ... a kind of social-control mechanism."* They are incapable of treating other humans as thinking, feeling beings.

So, combine a powerful animalistic instinctive nature with the spiritual, creative, active, dynamic, adventurous brain of the Cro-Magnon and what do you have? One has only to imagine the mixture of the non-creative, almost parasitic Neanderthal personality with the dynamic, creative Cro-Magnon to get an image of the aggressive, dominating authoritarian personality that is devoid of creativity, has no ability to conceive of time and space, and functions totally opportunistically – a hungry predator, though his hunger has new modes of expression and fulfillment, even if it can always be boiled down to flesh – to eat or mate with, or both: pure materialism. Perhaps that is the real 'Neanderthal Legacy'?

That is not to say that there is a single authoritarian or 'psychopath gene'. If you read Barbara Oakley's book *Evil Genes*, you realize that there are different systems involved in what we term Cluster-B personality disorders, including psychopathy. And each of these systems is controlled by different sets of genes. That would only be natural if we are talking about an event of mixing that took place a very, very long time ago. A neuropsych specialist of my acquaintance has proposed the following:

> The population of each band of Neanderthals was probably very small and was focused on survival in a rough environment. So I should qualify and say Neanderthals were psychopath-like, since the behavioral traits described certainly sound like it. Something was lacking in Neanderthals however, the intelligence and adaptability to manipulate in large groups. I imagine that any significant population comprised only of essential psychopaths would be evolutionarily unstable. Maybe that is why they died out? They were too psychopath-like but lacked adaptable traits.
>
> The survival of Neanderthal genes including those that make one prone to violence, aggression and lack of empathy would then require an infusion of genes from a more evolutionary stable population – one with traits of higher intelligence and adaptability in its genetic pool. This is what would seem more conducive to essential psychopathy as we understand it, hence it begins with mixing. Before that it was just a seed comprising aggression and lack of empathy but that is not all that makes an essential psychopath. The intelligence to do something with those traits, i.e., the ability to manipulate in various social situations is more like it.

According to Richard Bloom in *Evolutionary Psychology of Violence* – a book that I do not necessarily agree with in toto – even without an individual's motivation to spread their genes in the short term, groups do so in the long term. It would be even more imperative for small bands of aggressive types dealing with environmental disruption and population decline to rape females from a more stable and increasing population. It's probably why rape is more rampant during invasions and general upheaval. There is numerous documentation of primitive societies that raided for women. In essence, invaders would be more motivated to rape females precisely because of the increased probability that they may not survive the battle. Their genes are more likely to do so in offspring created by the deed.

That is to say, it was the mixing that created a problem amongst the Cro-Magnon population. It was a hybrid, not necessarily a mutation. Though, certainly, there can also be mutations across the spectrum of types due to epigenetic factors, not to mention other mutation-causing events, including cosmic factors. Some of the research suggests that this mixing took place in the Middle East between Neanderthal and the more gracile type of modern human. From there it could have spread to the more northern Cro-Magnon population. This, of course, suggests that the pure Cro-Magnon type may actually have come from the Fifth Planet physically rather than being the product of a rapid mutation event. There are so many factors to research it is really a shame that the evolutionists control science and prevent us from openly seeking solutions to the problems of our world.

In a very interesting paper titled, 'Neanderthal Settlement Patterns in Crimea: A landscape approach' [47], the author basically sets out the evidence that Neanderthals were so dumb they couldn't navigate in a landscape that didn't have clear, unusual features that could be seen from a distance, to help them to keep a marker in sight to find their way home when out foraging. It is suggested that this is why they never strayed far from their home area, never settled in areas that were not 'legible', and why their society was so static. The author demonstrates that Neanderthal camps were located next to striking landscape features, which suggests that those locations were chosen because they could be easily found. This suggests also that Neanderthals, after 200,000 years, still couldn't navigate by the sun, moon or stars.

I think one of the most useful books to read in order to really get a handle on how they may have used their minds is Steven Mithen's *The Prehistory of the Mind*. It is a fascinating book even if it is written by an evolutionist determined to persuade the reader to his view. But, what he does include in the way of cognitive science is excellent material to think about. There is a good bit of space devoted to analyzing the societies of other primates such as chimps and large apes, in order to compare the workings of their minds with man and to

---

[47] Ariane Burke, *Journal of Anthropological Archaeology*, 25:4, December 2006, pp. 510-523.

theorize how human thinking may have developed. He then goes on to discuss the Neanderthal question and it seems obvious that the Neanderthal society was rather similar to that of chimps – sometimes disturbingly so. There may be more to the belief of the evolutionists that they are descended from apes than is apparent on the surface. For them, it may be literally true, but not necessarily true for everyone. It all depends on the roll of the DNA dice.

There is a body of evidence attesting to mutations occurring at the times of cometary events collected in *The Diluvian Impact* by Heinrich Koch and *Man and Impact in the Americas* by E. P. Grondine. This evidence is in the form of legends of mutants that began to appear after impact events. Most of these legends have to do with alterations in body size (either giants or dwarves) and raging cannibalism. The two seem to have gone together. Perhaps they were really referring to periods when there were DNA polluting events, and this could include events between Neanderthal and Cro-Magnon or other hominids. Neanderthals were very short, Cro-Magnon man was very tall, so a hybrid of the two could be either/or, and both physical appearances could carry the personality of the Neanderthal with the intelligence and creative drive of the Cro-Magnon. A cannibal artist; not a pleasant thought.

What also strikes me as curious is the fact that redheaded individuals are noted to have been the particular objects of sacrifice in certain South American art. It is now known from the genetic mapping that Neanderthals were redheaded and pale skinned. Were the natives there encountering some of these hybrids and dispatching them? Or, worse, were some of these hybrids rising to power and teaching ordinary people their flesh-flaying, cannibalistic ways? This is, of course, assuming that much of the history of the Americas is much older than is presently allowed in the evolutionary scheme. Regarding that, Richard Firestone et al. makes an extremely interesting series of comments in *The Cycle of Cosmic Catastrophes*:

> Our early ancestors had only type O blood. Around the time of the supernova [40,000 years ago], mutations most likely occurred, creating blood types A and B. Types A and B blood are from dominant genes, so they spread through the population and became more common. ...
>
> DNA evidence suggests that B type blood probably originated in Central Asia or Africa, where the percentage is uniformly highest. Because the percentage is still very low in Australia and the Americas, it seems unlikely that it originated in either of those two places. Some geneticists conclude that type B is the youngest blood type, which appeared no earlier than 15,000 years ago and later than 45,000 years ago, and if so, this distribution seems inconsistent with early Americans originating in Asia and traveling across the Bering land bridge. If they had done so, there would be a lot more type B in the Americas. ...

For type A blood, the picture is more complicated, with apparent origins in Europe, Canada, and Australia. Again, there is little evidence that type A spread from Asia to the Americas. Instead, paradoxically, it appears to have arrived in the Americas from Europe long before Columbus did.

Is it possible that the Indians came from Europe? That idea seems far-fetched according to traditional views, and yet, according to Dennis Stanford, of the Smithsonian and Bruce Bradley (2000–2002), there is intriguing evidence connecting Clovis flint-knapping technology to the Solutrean flint technology in Spain at the end of the Ice Age. In addition, Clovis points are very unlike flint points from Asia, their supposed land of origin. Since blood types show a connection with Europe, perhaps there is one. Rather than Asians, maybe the Solutreans really discovered the New World – or perhaps others did, because, remarkably, recent studies of early South American skulls suggest aboriginal or African origins. …

Although type O blood is common everywhere, it is nearly universal among natives of South and Central America, and much more common in North American than in Asia or Europe. If people populated the Americas from Asia at the end of the Ice Age after types A and B arose, those people neglected to bring their normal distribution of blood types with them.

Another blood-typing system has been used to demonstrate the Asian origin of Native Americans. Called Diego, it evolved recently as a mutation, and all Africans, Europeans, East Indians, Australian Aborigines, and Polynesians are Diego-negative. East Asians and Native Americans are the only people that are Diego-positive. But Diego-positive is more common among Native Americans than among East Asians, raising the question of who got these genes first. From blood types alone, a case can be made that the oldest indigenous people are the Native Americans! (Firestone *et al.* 2006, 238–240)

The above might be seen as just confusing the matter, but if you will read Allan & Delair's book, *Cataclysm*, it will become clearer.

Back to Neanderthals: it is curious to me to note that they also had long 'bread loaf-shaped' heads with a 'bun' at the back. The 'mark of Cain', perhaps? It makes one wonder about the Ica skulls from South America as well as skulls such as those of Nefertiti and her children and how that may relate to the institution of monotheism. Periods of environmental stress enabled pathological types to rise to the top and they used religion to fulfill their bloody fantasies. I really do think that the psychopath – the extreme form of the authoritarian personality – derives enjoyment out of human suffering. Andrew Lobaczewski, the author of *Political Ponerology*, writes that *"Natural human reactions … strike the psychopath as strange, interesting, and even comical. They therefore observe us … They become experts in our weaknesses and sometimes effect heartless experiments."*

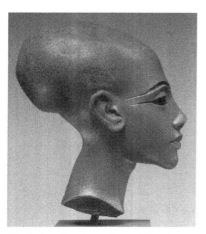

Head of an Egyptian princess. Was she one of the 'elite'?

One then has to consider such things as cranial deformation and circumcision. It is clear when you read studies of this sort of thing that many of the cases cited in the literature (Nefertiti and her children, for example) were not artificial deformation, but natural, and maybe this bizarre, dolichocephalic head with the extreme upward/backward extension was the biblical 'Mark of Cain' – the murderer. It also strikes me that, since there is an association between cranial deformation and circumcision, circumcision was also done in 'imitation' of the 'new elite', these hybrid beings with no souls. And if so, what was their genital structure? Were Neanderthal genitals different in some way back then, something that has been lost in the many generations of genetic recombination since those ancient times – and circumcision was an attempt to make the normal male organ look like that of the Neanderthal hybrid? There is a condition called *aposthia* in which a boy is born with a missing foreskin. The Midrash of Ki-Tetze notes that Moses was born aposthic. Other sources tell us that Jacob, his son Gad, and David were also born aposthic. Apparently, David Levy, former Israeli Foreign Minister and member of Knesset, was born aposthic. Arye Avneri's authorized 1983 biography of Levy notes this: *"When David Levy was born … his mother Sima noticed at once that he was different from other baby boys. He had been born already circumcised, for the foreskin was entirely missing."*

In any event, at some point, the serpent entered Eden, probably in the form of psychopathology from which a group of elites coalesced and people were pressing their children's heads and mutilating their penises to make them look like the rulers or to 'please the god'. After studying this topic (and cranial deformation was widespread and still practiced as recently as 50 years ago in some places including, believe it or not, Europe), I began to understand how humans get so twisted and wonder how any normal human being, subjected to those processes as infants, ever grew up sane at all. If they weren't personality disordered to begin with, they could become that way after being tied down with their heads compressed for the first two years of life, plus the genital mutilation factor (which is a lot worse for women).[48]

---

[48] See *Saharasia* by James DeMeo and *Artificial Cranial Deformation* by Eric J. Dingwall.

One point that needs to be addressed is the fact that a number of revisionist history authors have created some really bizarre fantasies about Neanderthals being gentle flower children decimated by the evil, technologically superior Cro-Magnon. These fairy tales are based on sensationalistic journalism. Let me just say this: reports of Neanderthal 'humanity' are greatly exaggerated. When you study the original reports carefully, it is pretty clear that Neanderthals had no beliefs or rituals and 'taking care of the elderly and infirm' is possible in one or two very late contexts probably due to imitation as a result of exposure to Cro-Magnon man.

In contrast to the stable, homogeneous society of Cro-Magnon man, social anthropologists point out the astonishing diversity of modern human societies as evidence for the plasticity of human psychology. Different societies have different standards, mores, religious beliefs, customs, rituals, and so much more. They then wonder how is this diversity even possible at all? The fact that human beings, who are said to be all members of the same species, can come up with so many different ways to order their lives, is extraordinary. No other species has this variability. One of the things this implies is that a species capable of this diversity across spatial boundaries can also be capable of great diversity over temporal boundaries. That is, they can change and that can mean growth or regression.

Another major point here is this: obviously, in general, human beings are not genetically programmed to be members of one or another specific social group, though there may be some genetic implications here (especially when you consider a genetic mix between Cro-Magnon and Neanderthal). What has been noted is that a child born into one social order can be raised from infancy in another social order and learn to function perfectly in that adopted order.

Each social group creates its own constraints and imposes them on its members. This means that there is a lot of diversity to be noted between different social groups, but *within any one group, diversity is not tolerated very well.* (Remember the authoritarian personality traits listed earlier?) If you move from being a member of one group to another group, you must change all your thinking, adapt to a different set of constraints that are imposed on you by the group you live among. Of course, some people can move from the social group they are born in to another quite easily because they are born with the mindset that is normal for that group. Anthropologists don't generally stay up to date on cognitive science findings.

What this means is that there is a genetic potential for diversification regarding thought and conduct born into every single human being, a potential that permits growth in any direction whatsoever, that can be inhibited and/or shaped by society. The question is, *how?*

One theory of human society is that of the 'social contract' which posits that a group of individuals get together and draw up an agreement to their mutual advantage by which they will all abide, and a 'society' is thus formed. The problem with this theory is that it is circular. It presupposes the very thing it purports to explain already exists: that human beings are already constrained by some values that allow them to get together to draw up this alleged contract. Such a group must already be able to conceptualize a situation in the future where they will benefit from being bound to these other people in a contract. (Keep in mind that this is an evolutionist's thoughts I am describing here.) Ernest Gellner outlines the basic theory of anthropology regarding how societies are formed. He writes:

> The way in which you restrain people from doing a wide variety of things, not compatible with the social order of which they are members, is that you subject them to ritual. The process is simple: you make them dance round a totem pole until they are wild with excitement and become jellies in the hysteria of collective frenzy; you enhance their emotional state by any device, by all the locally available audio-visual aids, drugs, dance, music and so on; and once they are really high, you stamp upon their minds the type of concept or notion to which they subsequently become enslaved.
>
> Next morning, the savage wakes up with a bad hangover and a deeply internalized concept. The idea is that the central feature of religion is ritual, and the central role of ritual is the endowment of individuals with compulsive concepts which simultaneously define their social and natural world and restrain and control their perceptions and comportment, in mutually reinforcing ways. These deeply internalized notions henceforth oblige them to act within the range of prescribed limits. Each concept has a normative binding content, as well as a kind of organizational descriptive content. The conceptual system maps out social order and required conduct, and inhibits inclinations to thought or conduct which would transgress its limits.
>
> ... I can see no other explanation concerning how social and conceptual order and homogeneity are maintained within societies which, at the same time, are so astonishingly diverse when compared with each other. One species has somehow escaped the authority of nature, and is no longer genetically programmed to remain within a relatively narrow range of conduct, so it needs new constraints.
>
> The fantastic range of genetically possible conduct is constrained in any one particular herd, and obliged to respect socially marked bounds. This can only be achieved by means of conceptual constraint, and that in turn must somehow be instilled. Somehow, *semantic, culturally transmitted limits are imposed on men.* [emphasis added] (Gellner 1995, 30–31)

While I have great admiration for Gellner, I have to point out that this theory of how to control human beings was understood in pretty much this way many thousands of years ago. In the course of my reading, I once came across a passage translated from an archaeological dig – a Hittite tablet – where the king wrote that the priesthood needed the king to establish their religious authority and the king needed the priests to establish his right to rule. This control comes sharply into view in the falsification of history. History itself becomes part of the control. After all, control of daily information is just history in the making. As to how this process works on the individual level, a passage in Barbara Oakley's *Evil Genes* describes what 'dancing around the totem pole with one's social group' does to the human brain – including scientists and creationists, both of whom have very strong attachments to their belief systems:

A recent imaging study by psychologist Drew Westen and his colleagues at Emory University provides firm support for the existence of emotional reasoning. Just prior to the 2004 Bush-Kerry presidential elections, two groups of subjects were recruited – fifteen ardent Democrats and fifteen ardent Republicans. Each was presented with conflicting and seemingly damaging statements about their candidate, as well as about more neutral targets such as actor Tom Hanks (who, it appears, is a likable guy for people of all political persuasions). Unsurprisingly, when the participants were asked to draw a logical conclusion about a candidate from the other – "wrong" – political party, the participants found a way to arrive at a conclusion that made the candidate look bad, even though logic should have mitigated the particular circumstances and allowed them to reach a different conclusion. Here's where it gets interesting.

When this "emote control" began to occur, parts of the brain normally involved in reasoning were not activated. Instead, a constellation of activations occurred in the same areas of the brain where punishment, pain, and negative emotions are experienced (that is, in the left insula, lateral frontal cortex, and ventromedial prefrontal cortex). *Once a way was found to ignore information that could not be rationally discounted, the neural punishment areas turned off, and the participant received a blast of activation in the circuits involving rewards – akin to the high an addict receives when getting his fix.*

In essence, the participants were not about to let facts get in the way of their hot-button decision making and quick buzz of reward. "None of the circuits involved in conscious reasoning were particularly engaged," says Westen. "Essentially, it appears as if partisans twirl the cognitive kaleidoscope until they get the conclusions they want, and then they get massively reinforced for it, with the elimination of negative emotional states and activation of positive ones" …

Ultimately, Westen and his colleagues believe that "emotionally biased reasoning leads to the 'stamping in' or reinforcement of a defensive belief, associating the participant's 'revisionist' account of the data with positive emotion or relief and elimination of distress. The result is that partisan beliefs are calcified, and the person can learn very little from new data," Westen says. Westen's remarkable study showed that neural information processing related to what he terms "motivated reasoning" ... appears to be qualitatively different from reasoning when a person has no strong emotional stake in the conclusions to be reached.

The study is thus the first to describe the neural processes that underlie political judgment and decision making, as well as to describe processes involving emote control, psychological defense, confirmatory bias, and some forms of cognitive dissonance. The significance of these findings ranges beyond the study of politics: "Everyone from executives and judges to scientists and politicians may reason to emotionally biased judgments when they have a vested interest in how to interpret 'the facts,'" according to Westen. (Oakley 2008, 189–190)

In short, a human brain that is calcified in a belief system – believing it because their social group, peers, family, also believe it – literally experiences pain if they try to open their mind and think in a limitless and unbiased way. The same thing must happen to a person trying to understand the different social constraints that are accepted as perfectly normal in a different cultural group. The implications are truly enormous. It can also explain why science is so corrupt. This is, essentially, a look inside the brain of the authoritarian personality.

But there is more to this than meets the eye. John Schumaker, in his book *The Corruption of Reality*, points out that human beings seem to come hardwired with a need to dissociate. I think that this is actually a need to make contact with the higher self – the field of information, if you like, that is transduced by our DNA. Some of us have DNA that connects us to the creative source, inherited from our Cro-Magnon ancestors, and some of us may have DNA that connects us to our Neanderthal ancestors – just a roll of the DNA dice, for the most part. In the Neanderthal-dominant individual (taking that as an hypothesis), the brain's ability to dissociate can simply be a normal state of being – a brain with no overseer or fully human consciousness enabling the coordination of the various parts and functions of the brain. It can also result in other mental disorders. In the individual with Cro-Magnon/ fully human DNA connection, the brain's capacity to dissociate has the potential to be utilized in a completely different way: a means of accessing and connecting to archetypal realms, realms of pure consciousness.

In any event, it seems clear from the evidence that Schumaker presents, that it is a hardwired function in all human-type beings that is just waiting to be taken advantage of by any snake-oil salesman that comes along. It is also abundantly evident that those who do not utilize this ability of the brain – those who suppress it – suffer from a myriad of physical disorders.

There is an epidemic in the world today that doesn't get nearly enough attention: the epidemic of stress. According to the latest statistics, human beings are 100 times more likely to have significant emotional/mental problems than people born a hundred years ago. Adult rates of depression and anxiety have tripled just since 1990. Over 80% of people who go to a doctor with physical problems also complain of excessive stress. These problems are increasing so fast that, within the next ten years or so, the primary causes of early death and disability will be stress, outranking most diseases, accidents and violence. So, something serious is going on and it is *really* bad. My guess is that it is due to the psychopathic (*aka* Neanderthal-dominant) effort to totally eliminate any concepts of spirituality from our reality. This is, in essence, what is at the root of evolutionary theory and why it dominates science today. Science has been taken over by Neanderthal-dominant individuals, human-looking beings who effectively do *not* have a spiritual nature though they may certainly have the higher brain functions with dead-end spirit connections; no possibility of developing conscience.

Lobaczewski points out that psychopaths naturally adopt a materialist, evolutionary worldview (because this is all they can conceive of). He remarks:

> Specialists in the areas of psychology and psychopathology would find an analysis of this system of prohibitions and recommendations [in scientific fields] to be highly interesting. This makes it possible to realize that this may be one of the roads via which we can reach the crux of the matter or the nature of this macrosocial phenomenon. The prohibitions engulf depth psychology, the analysis of the human instinctive substratum, together with analysis of dreams.

Even mating psychology is embargoed, lest women learn to avoid psychopathic males. Also forbidden is the study of psychopathology (psychopathy in particular). As he observes, any well-elaborated understanding of psychopathy would inevitably lead to a diagnosis of the system itself, an outcome that the pathological elites can't allow. So *"a purposeful and conscious system of control, terror, and diversion is thus set to work … The abuse of psychiatry for purposes we already know thus derives from the very nature of pathocracy as a macrosocial psychopathological phenomenon."*

This brings us to an inkling of what may have been at the root of the Cro-Magnon religion, the thing that made them peaceful, cohesive, creative, and homogeneous without force or domination. The solutions to the problems we deal with can suggest how they dealt with it. So, we notice that people need to know how to 1) deal with their stress in a healthy way, i.e., using dissociative techniques positively to focus on the non-material realm so as to enhance the ability to be objective when focused on the material world; 2) use dissociative techniques to fuse the many aspects of the conscious mind while, at the same time, making direct contact with the higher self.

The fact that certain techniques actually work to accomplish these things indicates to us the means by which humanity was taken over and led down the primrose path to destruction. Those true and workable techniques (which work because of the hardwiring of the brain) were utilized in a predatory way by the Neanderthal-hybrid snake-oilers to line their own pockets, to inculcate into people ideas and beliefs that are patently false and have no relationship to reality. They have been used to turn large groups of people against other large groups of people for the personal agendas of the snake-oil types and their cronies (the marriage between religion and politics is ancient), to foment wars, pogroms, persecutions, and so on.

It has always worked because that is the way the human brain is hardwired. And it is hard-wired because that it the way humans evolved. And they evolved that way because, clearly, for a very long period of time in our evolutionary history, it was advantageous to us to have this ability, which has now been used against us by pathological types. Lobaczewski is rare among psychologists in that he accepts this 'spiritual' hardwiring as natural, albeit undeveloped in most humans.

Given the above, it's not difficult to see how the corruption of science plays itself out, and what it means for humanity. Psychopaths, once in positions of power and influence, forbid areas of science that they know are dangerous to their position. It's a natural progression to apply this to the study of consciousness. They promote their own inner landscape (materialism) via science and project it onto humanity at large, effectively blocking the means by which they can be identified as abnormal. By creating such a semantic barrier, they can inhibit and shape our 'genetic potential for diversification of thought and conduct' in the direction of their choice. We lose the necessary tools with which to discover the true origins of their pathology, and the true potentials within normal humanity. By denying the existence of an ordering principle of consciousness, they deny the existence of any potential order to which we can strive. They block our return to the Golden Age.

So, how did they do it? How did pathological hybrids insinuate themselves into a stable, peaceful, homogeneous society that had been that way for literally thousands of years? I suggest that it was due to sex, due to some of these hybrids utilizing their predatory skills to spread their genes among the Cro-Magnon population.

This is a problem that has exercised me to no end. It's easy enough to just say that the hybrids went around raping women and spread their genes that way; certainly, some of that occurred. But when you consider the mixture of genetic types, you begin to realize that there were certainly some of them that put that predatory hunger for material goods and women to work in more creative ways due to the brainpower that had been acquired by the mixing. Not only that, infants born of such rapes would be liable to become victims of infanticide if there was not some compelling quality that prevented it. So, the obvious pathological types would have made themselves extinct and psychopathy would have selected for certain traits that are appealing and designed to ensure the survival of the infant.

In reading *Women Who Love Psychopaths* by Sandra Brown, I realized that the things that a psychopath does, the things that work in baiting, capturing, and bonding with women are obviously *caricatures* of things that ought to be manifested in positive ways. For example: a psychopath may use his eyes and words to entrance and bait a woman to his bed where he 'bonds' with her via 'super sex'. He uses tender, romantic words, gestures, promises, etc.

On the other hand, normal guys generally do not feel comfortable gazing into the eyes of their beloved, speaking romantic words, performing wildly romantic gestures and certainly, most men are sexually inhibited or downright juvenile in their sexual behavior. They also do not see sex as it ought to be seen, as one of the best opportunities they have in their daily lives for bonding with their partners. These negative and childish attitudes of men toward women are culturally imposed, mainly via religion – but it only starts there. To fully understand how twisted our society is regarding sex and sexuality, read Hervey Cleckley's *Caricature of Love*. It's a challenging book, but if you get through it, you will be glad you did – unless you are a psychopath, of course. And you'll see stark depictions of the very 'cannibal artist' hypothesized above.

A psychopath observes his prey, does all the things that he has learned will capture her, and then he bends her to his evil will. Why don't normal men observe their intended, not as prey, but as the object of devotion and giving? Why don't they learn everything about her, what she is, what she wants, what she needs, and then give it to her as an act of love?

Well, that's one thing that occurs to me. And the reason I bring it up is, as I said, because the interaction between the psychopath and his prey is a caricature of what seems to be part of the Cro-Magnon religious view of the world that prevailed in the Golden Age, and something that we have completely lost. If the stories about Sodom and Gomorrah are dim memories of the destruction of Atlantis conflated with later cometary impacts as I suspect, and if we consider the ancient stories linking the Deluge to some 'ritual fault' or sexual excess, we may now be getting some dim insight into what that might have been. It also gives all-new meaning to the story of the Nephilim:

And it came to pass, when men began to multiply on the face of the earth, and daughters were born unto them, That the sons of God saw the daughters of men that they were fair; and they took them wives of all which they chose ... There were giants in the earth in those days; and also after that, when the sons of God came in unto the daughters of men, and they bare children to them, the same became mighty men which were of old, men of renown. And God saw that the wickedness of man was great in the earth, and that every imagination of the thoughts of his heart was only evil continually. (Genesis 6:1–5, excerpts)

John van Seters thinks that the Bible was actually written very late and was heavily influenced by Greek mythology. Other experts agree. So is this excerpt from the bible just a take-off on the heroic tales of the Greeks, which are obviously about some seriously disturbed people? And are those ancient tales redactions of memories of cataclysmic times? (Again, see *The Diluvian Impact* and *Man and Impact in the Americas* for fairly comprehensive coverage of the actual stories and myths.)

Getting back to the caricature problem: As I said, I think that what psychopaths do is a caricature of what normal love between spiritual people might be like, how we evolved to interact with one another emotionally.

There are many people who feel acutely this lack of understanding in our world, who long for unity of purpose and to 'belong' to their 'own kind', so to say. This deep impulse is what the psychopathic religious and political leaders play upon to achieve their own entropic goals. This longing for union with another soul that understands and accepts one and loves unconditionally, who gives all and to whom one can give all is what is at the core of women who love psychopaths. And it is at the core of many people who become followers of religions and ideologies, and whose 'passion' is abused so that they become fanatics and 'radicals'.

In short, it seems to me that what psychopaths do works because they have observed women and know what to do to lure and capture them. And this works because these women have a certain 'something' inside them that is looking for a real love and they mistake the caricature for the real because they are ignorant of the facts of psychopathy. They don't realize that they have 'spiritual love binding-sites' that can be bound by a 'drug' (i.e., the psychopath) which does not act in the way the real neurochemical would act (i.e., the true spiritual love from a man who can give and receive true love). This idea raises a lot of issues, not the least of which is what genuinely spiritual men need to do to get over their hang-ups and learn to give on all levels: mentally, emotionally, physically and spiritually; and women who need to learn how to distinguish the true from the false.

Returning to our survey of what must have happened to lead humanity from the Golden Age into the type of society that can create so much cosmic discord that the celestial bodies themselves arrange for its extinction, we find Ernest Gellner suggesting that, because social constructs are little more than 'semantic limits' that are imposed on people from the outside, change is possible. But he notes that this must be cumulative, a sort of slow heating the frog in the pan of water.

> [This] also makes social change possible, change based not on any genetic transformation, but rather on cumulative development in a certain direction, consisting of a modification of the semantic rather than genetic system of constraints. ... The preservation of order is far more important for societies than the achievement of beneficial change ... most change is not at all beneficial; most of it would disrupt a social order without any corresponding advantage. (Gellner 1995, 31)

This is where we find the key to understand the corruption of science and thus, society and its understanding of the world that Zinoviev described. It seems that a new semantic construct has been in process since the Renaissance. Oddly enough, the Renaissance that gave birth to this new mode of thinking looked back almost 2000 years to the great minds of classical Greece and Rome for their mathematical, philosophical, and logical approaches. (Of course, Fomenko suggests, with strong evidence to back it up, that the 'ancient writings' of classical times were actually written during the Renaissance itself and were masterful forgeries.)

The 'evolutionary breakpoint' in the nineteenth century marked the culmination of a gradual shift in society from being dominated by religion to what was called 'rational thinking' and science. The new mindset was that the exercise of pure reason was the ultimate authority, not experience, authority or spiritual beliefs. This caused a shift in the way scholars saw themselves and the rest of humanity. This was, in general, a positive trend. This method laid the groundwork for a true scientific methodology and way of cognition. However, this attitude also fostered a very skeptical attitude toward prior belief systems and histories. Ancient civilizations were deemed primitive despite the obvious sophistication of the remains that were discovered, and stories of a lost Golden Age were rejected out of hand. Along came Darwin and his ideas became entrenched in all the sciences. So, it has been the steady application of materialistic evolutionary thinking that is behind the explanation of the order of the universe that prevails today. There are, undoubtedly, psychopaths in the woodpile here acting as the *éminence grise* behind science – the thing that controls most of our social constructs and institutions – because we certainly can't say that all scientists, or even most of them, are psychopathic. The profession itself excludes most psychopaths by virtue of the requirement for superior intellect. However, it can certainly include a great many members that are authoritarian in personality type.

There are other ways to induce changes and one of those is by the use of shocks – physical or emotional – to the human system which, if the shock is sufficient, more or less 'melts down' and becomes plastic enough to remold in a new configuration of social beliefs (or whatever is desired to be implanted as a control in the human brain). Pavlov called this 'Transmarginal Inhibition'. Social violence is a time-tested method used by elites to 'melt down' a population and make them amenable to changes in the environment that they would not previously have accepted. However, there is a problem here: the elites cannot do this openly against their own population because it would not be tolerated; it would be seen as an in-group violation. And so, the violence must be presented as that of an 'enemy' or a different social group.

Certainly, there are changes that occur in a society as a consequence of their organization being disrupted or destroyed by war – enemy attack – and in some cases it only strengthens the resolve of the attacked social group to hold more firmly to their social construct. It can also lead to shifts in values that incorporate the necessity of defense such as a peaceful people turning into valiant freedom fighters under the oppression of occupation.

Usually, occupying forces try to eliminate any individuals with a propensity to resist so that the task of re-shaping the society in question – 'pacifying' them, as they call it – is easier. Poland under Hitler and later under the Soviets is a case in point. The Germans considered the Poles to be an inferior race as they did other Slavic nations. So, according to the plan, the Poles were to be sorted according to strictly racist criteria. Those Poles with German ancestry were to be reclassified as ethnic Germans. The transformation of Poland into a German province was to be carried out over a short period of twenty-five or thirty years. Hence, no mercy was to be shown to this population. And, to guarantee the success of this fast despoliation, the intelligentsia was to be liquidated. *"It sounds cruel,"* Hitler reportedly told Hans Frank, *"but such is the law of life".*

12. As It Was In The Days Of Noë

Let's use some social anthropology to try to figure out what our Cro-Magnon society might have been like. Ernest Gellner points out the obvious: that the ability to be able to 'get along' with others, to avoid the chaos of the potential lack of constraint of the human being, to figure out how to utilize deeply internalized concepts, must have come first. He asks the question: which came first, this extraordinary ability to self-regulate by agreement with others and the wide variety of ways this can be done, or language which is necessary to even do these things? The two abilities are obviously interdependent. Once you have the possibility of an unlimited range of possible ways to behave, that behavior has to be restricted in order to get along with others, and there has to be some sort of system of signs that indicate the limits, if only in your own head. Language is a set of markers that delineate the boundaries of our conduct in our engagement with others; that is, language has replaced genetic limits on behavior.

This means that a number of things must have come together all at once: the possibility of a wide range of conduct, the means of signaling the optional and optimal ways of behaving between members of a group, and the presence of a system of inculcating those individuals into the system, i.e., ritual. This set of features must have arrived jointly, and that is a big problem for science when they try to use evolution as the explanation.

At the very same time, another set of features arrived. For most species, fire spells threat. So does direct eye contact and the opening of the mouth and the exposure of teeth. Combine these elements with the placing of food between a group of individuals other than parent and infant, and you have a recipe for conflict and violence in any other species. But, clearly, at some point, some group of humans turned all of these signals around and transformed them into the very essence of humanity. This change in behavior is evidenced by central hearths with clear spaces around them where diners gather in a circle. In the sharing of food and the accompanying courtesies, humans exhibit their fundamental distinction and separation from the animal kingdom.

Of all the things that humans do, sharing food is the least 'evolutionary'. No other species followed this path of mutual sharing and support, which required being able to conceptualize space and time abstractly. What logic of nature, as we witness it in the animal kingdom, would follow such a trajectory so foreign to the natural world?

As it happens, the communal meal is an important symbol to almost all ancient religious ensembles. The sharing of food penetrates to the most fundamental differences between views about humanity and the world at large. Perhaps they bear witness to Cro-Magnon vs. Neanderthal? I think it was a Paleolithic ritual that was handed down as a clue to reveal who was and was not pathological in those ancient times. Just think of 'The Last Supper' and its possible significance. What was so very important about sharing a meal that it became the centerpiece of a religion? Was there something about it that was older than Christianity as we know it? Perhaps the 'last supper' was something like: 'eat together – share food – in remembrance of me – you are not animals, you are spiritual beings'. Over millennia, bread replaced the original sharing of meat around the fire, but the symbology was still the same. The fact is, it seems that the largest brains and most 'human' behavior originated among those who cooked and ate meat.

The point is, on this earth – or on the Fifth Planet or Mars – there must have been a beginning and anyone who wants to understand the human dilemma cannot avert their gaze from the question of how it all began. We know that human beings are alienated from themselves and each other, and we desperately need to understand why in the fullest sense, not just a hodgepodge of theoretical patches. Of course, this presupposes that man has an 'essence' to begin with.

DaVinci's 'The Last Supper' - an example of the importance placed on communal eating.

Paleontology tries to discover the steps that would lead animals to fully human behavior. Maybe it isn't steps; maybe it is just simply the difference between what is animal-driven and what is truly human? But still, the research is useful to us in analyzing the differences.

Anthropology tries to discover by what steps can one proceed from a certain social order that is reasonably stable, imposed by ritual, to a society where humans can, of their own volition, commit themselves abstractly and at a distance, to specific behaviors that are not genetically imposed, i.e., the alleged 'social contract'.

Gellner believed that we live in such a society – the social contract kind. He felt that the ritually constrained society was neither normal nor acceptable. He failed to notice all the myriad ways in which dancing around the totem pole has been replaced by other equally effective ways of inculcating an individual into a group identity – the authoritarian dynamic that is probably driven by psychopaths and the semantic blocks they place on humanity via their promotion of certain sciences and embargo on others.

According to the mainstream scientific view, the Neolithic Revolution – the switch to agriculture – was one of the steps that led us to where we are today. This event involved the development of a system for the production and storage of food. Apparently, a human society already highly diversified and in the process of changing over to growing and storing food was already – according to Gellner and others – a 'ritually restrained society'.

> But it [agriculture] was also a tremendous trap. The main consequence of the adoption of food production and storage was the pervasiveness of political domination. A saying is attributed to the prophet Muhammad which affirms that subjection enters the house with the plough. This is profoundly true. The moment there is a surplus and storage, coercion becomes socially inevitable, having previously been optional. A surplus has to be defended. It also has to be divided. No principle of division is either self-justifying or self-enforcing: it has to be enforced by some means and by someone.
>
> This consideration, jointly with the simple principle of pre-emptive violence, which asserts that you should be the first to do unto them that which they will do unto you if they get the chance, inescapably turns people into rivals. Though violence and coercion were not absent from pre-agrarian society, they were contingent. They were not, so to speak, necessarily built into it. But they are necessarily built into agrarian society ...
>
> The need for production and defense also impels agrarian society to value offspring, which means that, for familiar Malthusian reasons, their populations frequently come close to the danger point ... The members of agrarian societies know the conditions they are in, and they do not wait for disaster to strike. They

organize in such a way as to protect themselves, if possible, from being at the end of the queue.

So, by and large, agrarian society is authoritarian and strongly prone to domination. It is made up of a system of protected, defended storehouses, with differential and protected access. Discipline is imposed, not so much by constant direct violence, but by enforced differential access to the storehouses. Coercion does not only underwrite the place in the queue; the threat of demotion, the hope of promotion in the queue also underwrites discipline. Hence coercion can generally be indirect. The naked sword is only used against those who defy the queuemasters altogether ...

... the overwhelming majority of agrarian societies are really systems of violently enforced surplus storage and surplus protection ... Political centralization generally, though not universally, follows surplus production and storage. ... A formalized machinery of enforcement supplements or partly replaces ritual. (Gellner 1995, 33–34)

The next step is, of course, writing because writing is necessary to keep track of what is stored, who is in the queue in what rank, and so on. Writing then leads to something else interesting: the codification of meanings and the storage of these codifications, generally dogmatic ideas. This means that ideas can be shared across time and space. If the doctrine is centralized and endowed with a single apex/origin, i.e., an exclusive and jealous god, this can have very bad consequences across time and space. This then leads back to coercive semantic techniques with a kick: it is widespread and centralized with a very small ruling elite. There are the specialists – priests – who legitimate the beliefs, and the specialists of violence that enforce it. Their pre-eminence testifies to the fact that the maintenance of social discipline is a problem and is seldom attainable without them.

Society constituted on the principles of the agrarian surplus-and-storage way of life cannot manage without the help of the elite and once they have acquired an elite, it is very, very hard to resist their growing demands, and even harder to get rid of them. Moreover, an agrarian society based on accumulating, storing, and dispensing material goods is easily converted to an industrial society, and so it is. Anthropologically and sociologically, that's the short version of how it happened. Somehow, man with full conceptual abilities leapt upon the stage of history and started dancing around the totem pole and really hasn't stopped.

So, what do I think that society was like in the Golden Age? Again, the answer can be found in social anthropology.

According to the most recent archaeological and paleontological research, the very oldest form of religion that can be identified with Cro-Magnon man was the worship of the Celestial Mother Goddess. The sculpted figures of the Paleolithic era are, in a way, directly connected to contemporary images of the Virgin Mary who would, in the terms of my previous chapters, be a class-A 'witch', or Holy Person. Thus it seems that there has been a continuous transmission of this striking concept throughout human history. The implicit message in the image of the goddess is a vision of all life as a living unity. The feminine image is like a lens that focuses our perception of the universe as a sacred whole, alive and giving, and we on earth are the children of the cosmos. In this context, you could say that the Holy Grail is the Goddess religion – using religion in its original sense, something that binds people together. The earth and all creation were the same substance as the goddess and the divine was immanent as creation.

This is what we have lost. Nowhere in our present Dark Age is the goddess

The Venus of Wilendorf – an example of a Paleolithic era mother goddess.

image – the Holy Grail lens through which to view our world – to be found. The image of the Virgin Mary is not one of Queen of the Earth; she is, significantly, only 'Queen of Heaven'. Since mythic images ultimately govern a culture, and ours is a culture that has debased the feminine for many thousands of years, we find here the root of the horror of our reality: that we have desacralized Nature. The earth is not experienced as a living being; instead, it is polluted almost to the point of destruction. 'Pollution' originally meant the profaning of the sacred, so the term is quite apt.

Here is where we must consider the organizing principle of Cro-Magnon humans no matter where they came from. At some point, there was an impulse to form societies, whether this was due to spontaneous generation of DNA or long, slow, painful evolution somewhere, some time, somehow. It doesn't matter which it was, the principle is still the same. Clearly, something special happened for animals to become human, something that differentiated man from his proto-human ancestors (and thus far, evolution, *as it is formulated*, has been abysmally inadequate to explain this).

We can examine the society of Neanderthal to see what Cro-Magnon was distinctly not. Proto-human gangs – Neanderthals – were small, cohesive bands dominated by violence. The strong dominated the weak, but they needed to be cohesive to survive. It was rude and vulgar competition and conflict for resources – both food and females. The cohesion was enforced, a sort of rule by thuggery. Such an environment is not conducive to any kind of innovation such as the creation and development of tools or the ideas that would lead to tools. Studying the history of Neanderthal man suggests that this was, indeed, life in the Neanderthal camp. Two hundred thousand years of the same old recycled tool kits with never an innovation.

Innovative individuals, as we know them in our own societies, are very often not physically strong or aggressive. Perhaps they think because they are less able to act in the world physically or perhaps they become less aggressive because they think so much. Either way, such individuals would not survive in a rule-by-thugs society because the thugs would not only steal their innovations from them, (and be unable to reproduce them themselves) but would not feed them either, so such a dog-eat-dog society is not a place where creativity and innovation can flourish.

Cro-Magnon societies included communal hearths, communal activities, and the sites show that areas were set aside where groups of people performed different tasks for the good of the group as a whole. And thus it is suggested by Marxist social anthropologist, Yuri Semenov, that communism (the real kind as practiced by the early Christians, not the Soviet version which was totalitarianism with the name of 'communism' attached) was something practiced by the earliest true humans – the thing that made them human, or at least expressed their humanity – and something which was the essence of life itself.

In an egalitarian, sharing, communistic community where each individual is taken care of and gives back what he is capable of giving, innovators can flourish. The toolmaker that cannot hunt gets fed in exchange for making the hunting tools, to give a very simplistic example. The bottom line is this: without a communistic system in operation, there would be no advances for humanity at all. The problem is, we are inculcated into a psychopathic reality from birth and all our thinking processes (programs, if you like), develop as a way to survive in such a reality. Those programs include the early implantation of a competitive, dog-eat-dog, us-*vs.*-them, authoritarian mentality. We simply cannot imagine life in a different world where caring replaces competition, where no one ever asks 'am I my brother's keeper' because all understand from infancy that 'I am my brother'. This is because psychopaths *lack* these essential human qualities that foster social ties, sharing, bonding, caring (what Lobaczewski called *"natural syntonic responses"* that are a product of our human instinctive substratum).

Communism definitely explains how early humans could have survived and developed societies; we can see that it was a necessary condition, but we can't see how it came about. What would turn an animalistic creature that is only concerned with food, mating, possibly warmth, and little else into a gentle, caring strong man who looks after the weaker members of his society? Science cannot answer this question satisfactorily. Oh, yes, they use 'natural selection' as the god that endows humans with a collection of characteristics by killing off the ones that don't have them, but the same problem exists: a whole set of characteristics would have to have emerged fully formed on the stage of history at once. It is totally absurd to follow the evolutionary line of reasoning on this point where it is said that natural selection operated on bands of individuals rather than individuals because that already suggests that a band of proto-humans included members that acted in truly human ways. And keep in mind that I am in no way denying the workings of evolution in a general sense: it is clear that evolution is a big part of the processes of life, but there is something more to it than the strict materialistic version accommodates.

Agriculture goes hand in hand with the dominating, punishing god. From Babylonian mythology onward, (and undoubtedly before, but this is the earliest that is recorded), Nature began to be described as a chaotic force to be mastered, and the god took the role of conquering it and imposing order on the chaos. The goddess became almost exclusively associated with Nature and the god took on the quality of a countering 'spirit'– an opposition that had not previously existed. As a result, humanity and nature became polarized – in opposition to one another – and it is the Judeo-Christian mythology that incorporated the Babylonian myths, giving rise to our present-day afflictions.

This attitude is implicit in the assumption that the spiritual and physical worlds are different and opposed, and this has led to other splits in our thinking, divorcing mind from matter, soul from body, thinking from feeling, intellect from intuition, making things black and white, evil and good, and with no understanding of the third force: context.

The feminine principles of spontaneity, instinct, and intuition, have been lost as our guides to experience the sanctity and unity of all life. In Judeo-Christian thought, there is no feminine dimension and our culture is formed in the image of a masculine god who orders everything from outside and beyond creation where, of course, the Virgin Mary is under his thumb. It was only in 1950, by popular demand, that the Virgin Mary was finally declared 'Assumed into Heaven, Body and Soul'.

In other mythic constructs, the goddess did not go away entirely, even if she was debased; for example, Zeus (and other gods) 'married' the goddesses and they retained their places to some extent, presiding over fertility, childbirth, the home, and even spiritual transformation. Hebrew mythology – which was adopted by Christianity – however, forced the goddess completely underground; she became dragons: Leviathan and Behemoth or the evil Canaanite goddess, Astarte, and more abstractly, Yahweh's wisdom, Sophia, or his presence, Shekinah. Eve, the temptress, was human and susceptible to evil, and Adam gave her the name of the goddess: 'Eve, mother of all living'. This little twist was fatal and limiting to the feminine principle.

But the reality of the feminine principle is always there, lurking in the subconscious of man. The action of the goddess, denied in formal doctrine, acted implicitly and indirectly, unacknowledged, but persistent and often distorted. Attempting to eradicate the feminine principle was equivalent to trying to destroy fully half of humanity, not to mention half of the human psyche in all humans, men included.

Cro-Magnon came from somewhere, and something corrupted the world of Cro-Magnon. Whatever it was that led peaceful hunter-gatherer types to create a complex technological civilization that then took the peoples of the world into an abyss, may very well still be present in our own society. And it is doing its darndest to destroy the one thing that could save us: true science.

It was in the discoveries of science that the goddess first began to emerge again. And here, the image emerges as it was in the days of old: not a personalized being, but rather a vision of sacred life, a whole, in which all living beings participate and relate dynamically to one another. Physics demonstrated that in the subatomic universe, there was chaos that was waiting and willing to become real in a relationship with the observer, sort of like love between the lover and the beloved. The web of space and time is a characteristic image of the old goddess myths where the mother goddess once spun the cosmos as a web from her eternal womb via a spindle, whorls, to become the cosmic web of life. The Ocean of limitless energy is the image of all the mother goddesses who were born from the sea. But of course, even in physics, the goddess was assaulted with the splitting of the atom. In short, the predominant mythic image of the Dark Age – the god without a goddess – continues to perpetuate and support the oppositional and mechanical paradigm that science itself refutes. The effects on the minds of human beings are leading us toward unparalleled disaster.

The two essential aspects of the human spirit are still in opposition to one another and the human mind has lost its way. Mythic images have a profound effect on humans, individually and collectively, and this is clearly shown in depth psychology, which reveals how radically we are influenced and motivated by that which lies below the surface of our waking consciousness. It is also why it is one of the domains ridiculed and forbidden by 'official' (psychopathic) science.

Today, high technology is considered to be the hallmark of a high civilization. And so we judge the ancients by our standards. However, how have space travel, computers, digital animation or heart transplants improved our lives? How has a society based on the theory of evolution brought us peace and prosperity for all mankind? Today, more than ever, meaning and purpose are lacking even though we have everything in the way of technology that our hearts could desire. I had a powerful dream once where a voice came out of a light and told me that any technology that does not require equal input from a human being to be functional is entropic and leads to decay and disorder in the cosmos.

Our ancestors of the Paleolithic Golden Age appear to have had many technologies that we don't have. These technologies were subtle, working with natural energies, in harmony with the world and the people rather than seeking to overpower it.

Today, people value their possessions and perks of the good life as a sign of success. They have cars, houses, cash, stocks, holiday homes, and so forth, and wonder why the earth has turned against them.

We know from mythology that 'peace and plenty' was the hallmark of the Golden Age. It seems that the status of the individual was measured by knowledge and enlightenment; witness the honor accorded to the bard, the sage, and the shaman. Yes, they seem to have put their emphasis on spirituality and loving and honoring Nature and the cosmos, but they were certainly not devoid of material comforts. Archaeological studies show that they lived well, ate well, were interested in astronomy, mathematics, religion, and were capable of producing luxury goods that were equitably distributed. The quality and quantity of fine jewelry dating to the most ancient times is staggering. The comforts enjoyed by the Cro-Magnon peoples continually surprise archaeologists, and it is a certainty that they were surrounded by the most sophisticated and enthralling art ever known. There are numerous indicators that our Golden Age ancestors lived not only a rich and comfortable life; they also lived longer lives.

To draw this discussion to a close, I want to say that science – true science that is open enough to explore even our spiritual realities – is our only hope. It has only been via scientific research that we have found the *disjecta membra* of a lost civilization, a universal prehistoric cosmology that sustained a Golden Age for many thousands of years. We still have some of the ideas and stories about the nature and the origin of man that were formerly the birthright of people around the entire planet earth.

If we admit the possibility of agencies of consciousness involved in evolution, then we may also consider that these agencies can be numerous and different. Given the great diversity of Nature, we can suppose that there are

The destruction of Sodom and Gomorrah.

as great a variety of spiritual agencies as there have been forms of life on earth. Archangels, angels, and devils are familiar classical ideas that are today considered outmoded between the monotheistic gods of creationism and evolutionism. But the existence of hosts of spiritual agencies was the answer articulated by the ancients. Perhaps they were right about that as they have been shown to be right about so many other things?

So, here we stand, quite possibly on the threshold of a new age, perhaps even a new Golden Age. It is very difficult to really understand its true potential. Yes, it is very likely going to be an extinction event, but as usually happens in such events, some survive.

The ancient stories of survivors all have one thing in common: those who could see the signs and knew what was coming, were ready. As it was in the days of Noë...

Bibliography

Allan, D. S. and J. B. Delair. *Cataclysm! Compelling Evidence of a Cosmic Catastrophe in 9500 BC.* Bear & Co., 1997.

Arendt, Hannah. *The Origins of Totalitarianism.* Harcourt, 1976 [1951].

Baillie, Mike. *Exodus to Arthur: Catastrophic Encounters with Comets.* London: B.T. Batsford, 1999.

Baillie, Mike. *New Light on the Black Death: The Cosmic Connection.* Tempus, 2006.

Bobrowsky, Peter T. and Hans Rickman (eds). *Comet/Asteroid Impacts and Human Society: An Interdisciplinary Approach.* New York: Springer, 2007.

Brown, Sandra L. *Women Who Love Psychopaths.* 2nd edition. Mask Publishing, 2010.

Cheney, Edward P. *The Dawn of a New Era: 1250-1435.* 1936.

Cleckley, Hervey. *A Caricature of Love: A Discussion of Social, Psychiatric, and Literary Manifestations of Pathologic Sexuality.* Grande Prairie, AB: Red Pill Press, 2011.

Clube, Victor and Bill Napier. *The Cosmic Serpent: A Catastrophist View of Earth History.* London: Faber & Faber, 1982.

Clube, Victor and Bill Napier. *The Cosmic Winter.* Oxford: Basil Blackwell, 1989.

Davis, Paul K. *100 Decisive Battles from Ancient Times to the Present: The World's Major Battles and How They Shaped History.* Oxford: Oxford University Press, 1999.

DeMeo, James. *Saharasia: The 4000 BCE Origins of Child Abuse, Sex-Repression, Warfare and Social Violence, in the Deserts of the Old World.* Natural Energy Works, 2006.

Dingwall, Eric John. *Artificial Cranial Deformation: A Contribution to the Study of Ethnic Mutilations.* London: John Bale, Sons and Danielsson, 1931.

Donnelly, Igantius. *The Destruction of Atlantis: Ragnarok, or the Age of Fire and Gravel.* Dover, 2004 [1883].

Eliade, Mircea. *Shamanism: Archaic Techniques of Ecstasy*. Princeton University Press, 1964.

Firestone, Richard, Allen West and Simon Warwick-Smith. *The Cycle of Cosmic Catastrophes: How a Stone-Age Comet Changed the Course of World Culture*. Rochester, Vermont: Bear & Co., 2006.

Fix, William R. *The Bone Peddlers: Selling Evolution*. New York: Macmillan, 1984.

Fomenko, Anatoly T. *History: Fiction or Science?* Chronology 1. Delamere Publishing, 2006.

Garcia-Ballester, Luis; Roger French, Jon Arrizabalaga, Andrew Cunningham (eds). *Practical Medicine from Salerno to the Black Death*. Cambridge University Press, 1994.

Garnier, John. *The Worship of the Dead, or the Origin and Nature of Pagan Idolatry and Its Bearing Upon the Early History of Egypt and Babylonia*. London: Chapman & Hall, 1904.

Gehrels, Tom (ed). *Hazards Due to Comets and Asteroids*. University of Arizona Press, 1995.

Gellner, Ernest. *Anthropology and Politics: Revolutions in the Sacred Grove*. Wiley-Blackwell, 1995.

Gimbutas, Marija. *The Language of the Goddess*. New York: Harper & Row, 1989.

Grondine, E. P. *Man and Impact in the Americas*. 2005.

Hare, Robert. *Without Conscience: The Disturbing World of Psychopaths Among Us*. Guilford Press, 1999.

Hoyle, Fred. *The Origin of the Universe and The Origin of Religion*. Moyer Bell, 1997.

Klein, Naomi. *The Shock Doctrine*. New York: Picador, 2007.

Knight, Christopher and Robert Lomas. *Uriel's Machine*. Fair Winds Press, 2001.

Knight-Jadczyk, Laura. *The Secret History of the World and How To Get Out Alive*. Grande Prairie, AB: Red Pill Press, 2005.

Koch, Heinrich P. *The Diluvian Impact*. Peter Lang, 2000.

Kors, Alan Charles and Edward Peters. *Witchcraft in Europe, 1000–1700: A Documentary History*. University of Pennsylvania Press, 1972.

Leakey, Richard E. and Roger Lewin. *The Sixth Extinction: Patterns of Life and the Future of Humankind*. Anchor, 1996.

Lewis, John. *Rain of Iron and Ice: The Very Real Threat of Comet and Asteroid Bombardment*. Helix Books, 1997.

Lewis, John. *Comet and Asteroid Impact Hazards on a Populated Earth: Computer Modeling*. San Diego: Academic Press, 2000.

Lobaczewski, Andrew. *Political Ponerology: A Science on the Nature of Evil Adjusted for Political Purposes.* Grande Prairie, AB: Red Pill Press, 2006.

Mack, Burton. *A Myth of Innocence: Mark and Christian Origins.* Fortress Press, 1988.

McCanney, James M. *Planet-X, Comets and Earth Changes.* Minneapolis, MN: jmccanneyscience.com Press, 2007.

Mellars, Paul. *The Neanderthal Legacy: An Archaeological Perspective from Western Europe.* Princeton University Press, 1996.

Milton, Richard. *Shattering the Myths of Darwinism.* Inner Traditions, 1997.

Mithen, Steven. *The Prehistory of the Mind: The cognitive origins of art, religion and science.* Thames & Hudson, 1996.

Oakley, Barbara A. *Evil Genes: Why Rome Fell, Hitler Rose, Enron Failed, and My Sister Stole My Mother's Boyfriend.* Prometheus, 2008.

Russell, Jeffrey Burton. *A History of Witchcraft, Sorcerers, Heretics and Pagans.* London: Thames and Hudson, 1980.

Schechner, Sara J. *Comets and Popular Culture and the Birth of Modern Cosmology.* Princeton University Press, 1999.

Schumaker, John F. *The Corruption of Reality: A Unified Theory of Religion, Hypnosis, and Psychopathology.* Prometheus, 1995.

Shiller, Bryant M. *Origin of Life: The 5th Option.* Trafford Publishing, 2006.

Sprenger, Jacobus and Heinrich Kramer. *Malleus Maleficarum.* Tr. Montague Summers. London: Folio Society, 1968.

Trevor-Roper, Hugh. *The Crisis of the Seventeenth Century: Religion, the Reformation, and Social Change, and Other Essays.* Indianapolis: Liberty Fund, 2001 [1967].

30938013R00151

Made in the USA
Lexington, KY
23 March 2014